The vegetation types of Northeast Greenland

A phytosociological study based mainly on material left by Th. Sørensen from the 1931-35 expeditions

BENT FREDSKILD

BENT FREDSKILD: *The vegetation types of Northeast Greenland. Meddelelser om Grønland, Bioscience 49. Copenhagen, The Commission for Scientific Research in Greenland, 1998.*

Printed by Datagraf Auning as
Set in Minion

Scientific Editor:

Dr. Gert Steen Mogensen, University of Copenhagen, Botanical Museum, Gothersgade 130, DK-1123 Copenhagen K. Phone + 45 3532 2202, fax + 45 3532 2210, email gertsm@bot.ku.dk.

About the monographic series Meddelelser om Grønland

Meddelelser om Grønland, which is Danish for Monographs on Greenland, has published scientific results from all fields of research in Greenland since 1879. *Meddelelser om Grønland* is published by the Commission for Scientific Research in Greenland, Denmark. Since 1979 each publication is assigned to one of the three subseries:

- *Man & Society*
- *Geoscience*
- *Bioscience*

Bioscience invites papers that contribute significantly to studies of flora and fauna in Greenland and of ecological problems pertaining to all Greenland environments. Papers primarily concerned with other areas in the Arctic or Atlantic region may be accepted, if the work actually covers Greenland or is of direct importance to continued research in Greenland. Papers dealing with environmental problems and other borderline studies may be referred to any of the series *Geoscience, Bioscience or Man & Society* according to emphasis and editorial policy.

For more information and a list of publications, please visit the web site of the Danish Polar Center *http://www.dpc.dk*

All correspondence concerning this book or the series *Meddelelser om Grønland* (including orders) should be sent to:

The Commission for Scientific Research in Greenland
Danish Polar Center
Strandgade 100 H
DK-1401 Copenhagen
Denmark
tel +45 3288 0100
fax +45 3288 0101
email dpc@dpc.dk

Accepted January 1998
ISSN 0106-1054
ISBN 87-90369-28-9

Contents

The vegetation types of Northeast Greenland

A phytosociological study based mainly on material left by Th. Sørensen from the 1931-35 expeditions

ABSTRACT

Based on 347 vegetation analyses carried out in 1931-35 by Th. Sørensen in the 72°-74°N area, supplemented by 136 recent analyses from 74½°N and field observations northwards to 79°N, the vegetation types of Northeast Greenland are described. Further to phanerogams, the mosses and a few lichens are determined in 274 of the 347 analyses.

The most characteristic vegetation types present in this middle-arctic area are: 1. Dwarfshrub heaths dominated by *Cassiope tetragona, Salix arctica, Vaccinium uliginosum* ssp. *microphyllum* or *Betula nana*, depending on soil and duration of snow cover. 2. Grasslands, drying out during summer, characterised by *Arctagrostis latifolia, Carex bigelowii, C. misandra*, and *Eriophorum triste*. 3. Permanently wet fens with *Carex stans, Eriophorum scheuchzeri*, and *Arctagrostis latifolia*. 4. Snowbeds, the late ones characterised by *Phippsia algida*, the moderately late by *Salix herbacea*, and the few early herb-slope like snowbeds by *Trisetum spicatum* and *Erigeron humilis*. 5. Open, graminoid *Dryas*-heaths and fell-fields on dry soil with *Carex nardina, C. rupestris*, and *Kobresia myosuroides*. 6. Fell-fields with *Calamagrostis purpurascens* and *Carex supina* ssp. *spaniocarpa*. 7. Species rich communities on wet ground, covered by organic crust, characterised by *Koenigia islandica* and *Festuca hyperborea*. 8. Halophytic vegetations with *Puccinellia phryganodes* and *Carex subspathacea*.

Soil samples have been analyzed for nine factors for the major part of the 347 analyses. The plant to soil relationship is illustrated for the 109 most frequent phanerogam species, for 40 of these to all nine factors, for 69 only to pH and conductivity.

Sørensen arranged his analyses in 41 phytosociological groups according to his similarity coefficient - the "Sørensen index". These groups form the basis of the phytosociological classification according to the Braun-Blanquet system. The vegetation units are grouped into 13 associations, of which 7 are new, subdivided into a number of new, lower phytosociological units. These are grouped in the alliances Saxifrago-Ranunculion nivalis, Caricion atrofusco-saxatilis, Dryadion integrifoliae, Veronico-Poion glaucae, and Puccinellion phryganodis. Two synoptical tables summarizes the frequency of 135 phanerogams and 99 cryptogams, mainly mosses, in 42 phytosociological units.

Keywords: Greenland, middle-arctic vegetation, phytosociology, vegetation types, Sørensen index, syntaxon, Braun-Blanquet.

Bent Fredskild, Greenland Botanical Survey, Botanical Museum, University of Copenhagen, Gothersgade 130, DK-1123 Copenhagen K, Denmark.

1. Introduction

Thorvald Sørensen participated as botanist in Lauge Koch's "Three Year Expedition to Northeast Greenland 1931-34". From Ella Ø, which was his wintering base, he worked in 1931-32 with phytosociology on Ella Ø, Ymer Ø, Traill Ø, and Kap Hedlund (Figs 1-2). In 1933 he collected plants at a number of localities around Skærfjorden and on Kuhn Ø. In 1934-35 he wintered at Eskimonæs on Clavering Ø, working mainly with "Temperature relations and phenology of the Northeast Greenland flowering plants" which became the title of his thesis (Sørensen 1941). In August 1934 he finished the phytosociological work at some outer coast localities on Hold with Hope, and in 1937 he collected in the Scoresby Sund area.

During these years Sørensen developed the idea to treat his data statistically based on his concept of similarity of species contents. It has been told, that he realized it would not be possible to finish the work with 392 vegetation analyses in a foreseeable future as the most advanced facilities for calculating were logarithmic table and slide rule. Because of this, he published his method of establishing groups of equal amplitude in plant sociology based on similarity of species content - later known as the Sørensen index - on fewer samples, viz. 50 analyses of the vegetation on Danish plant commons (Sørensen 1948). However, he continued working on the Northeast Greenland material according to the published principles, leaving at his death in 1973 a

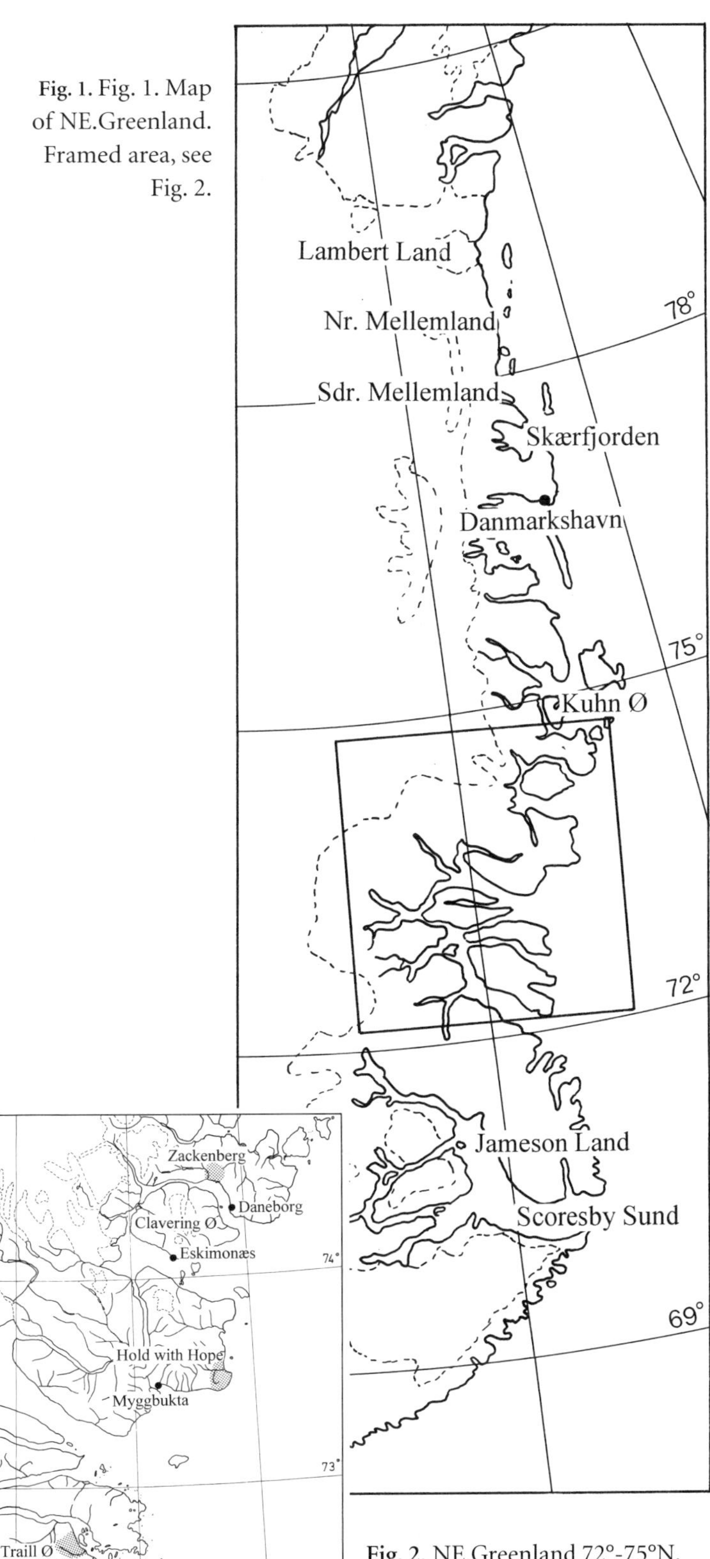

Fig. 1. Fig. 1. Map of NE.Greenland. Framed area, see Fig. 2.

Fig. 2. NE.Greenland 72°-75°N. The phytosociological investigations have been carried out in the six shaded areas.

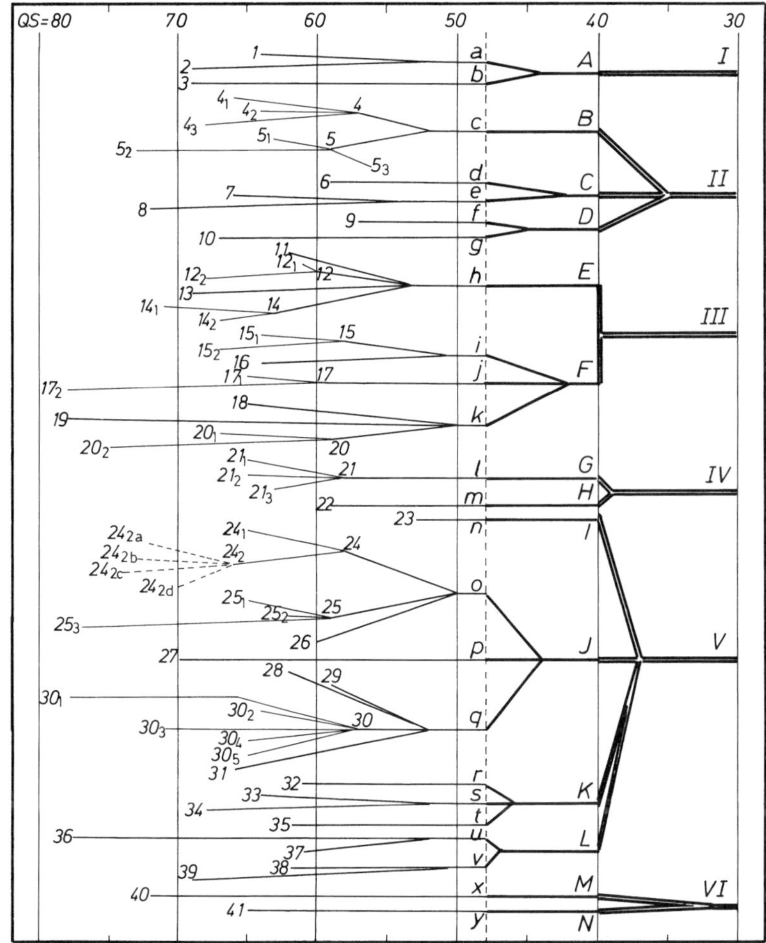

Fig. 3. Dendrogram based on the similarity quotient (QS) between the 41 groups in which Th. Sørensen arranged the vegetation types in his investigation areas.

dendrogram (Fig. 3), tables of the vegetation analyses within each group, a table showing four signatures of the frequency of 156 phanerogam taxa arranged in 62 groups and subgroups. Further, an undated summary (in Danish): "A summary account of an unfinished treatment on the plant communities of Northeast Greenland (72°-74°N) based on vegetation analyses, carried out 1931-33" (Sørensen s.a., 9 pp).

In the early 1960'ies 268 soil samples were analyzed at the Department of Plant Ecology of the University of Copenhagen, and mosses and other cryptogams from 274 vegetation analyses were determined by K. Holmen, who died in 1974. Only a minor part of the lichens were determined. On the initiative of one of the participants of the summer expedition in 1932, Mogens Køie, a group of botanists, including the author, attempted to publish this material, and much work on typing tables, drawing diagrams illustrating the soil analyses, etc. was carried out 1976-78; however, publication was postponed. In 1993 M. Køie handed over the material to the author.

The purpose of this paper is to present the summary of the results of the work by Sørensen, supplemented by the soil analyses and the most recent vegetation analyses by the author.

2. Investigation area

2.1. Geology

Broadly, the main geological belts run parallel to the outer coast of Northeast Greenland, with crystalline rocks at the head of fjords, sediments in middle fjord areas, and sediments and basalts at the outer coast (Fig. 4). In Sørensen's working area (Sørensen 1933, Fig. 2, in this paper termed the 73°N area), the peninsula Kap Hedlund in the interior is formed of gneiss. His middle district includes the isles of Ella Ø and Ymer Ø. The former of these is formed of calcareous sedimentary rocks, with the exception of one site (on which some of the analyses were made), intersected by crests of tillite, poor in lime. On Ymer Ø, the broad valleys on the middle part of the south side were sampled. The surrounding mountains are partly built up of calcareous sediments, partly of sandstone, whereas lower parts of the valleys are raised marine beds. In the outer district, an area on the northeast side of Traill Ø was sampled. It is formed of sandstone, and a basaltic plateau 500-700 m a.s.l. The other outer coast area sampled is the basaltic lowland on the southeastern corner of the peninsula Hold with Hope and a little island, Holland Ø, just outside.

In the Zackenberg valley, sampled by the author in 1992 and 1996, the north-south running river separates steep gneissic mountains to the west from more gently sloping sandstone mountains overlayed by basalt to the east, clearly reflected in the frequency of several species and plant communities. The valley bottom is formed partly of moraines deposited on meltwater plains, partly on raised marine deltaic beds.

2.2. Soil

In the Mestersvig area the most common soils are upland tundra soils, soils of the gelifluction slopes, and lithosols. Other soils here include all phases of arctic brown soils, found on c. 10 % of the area, mainly on moraines, delta remnants, kames, and raised beaches, and further

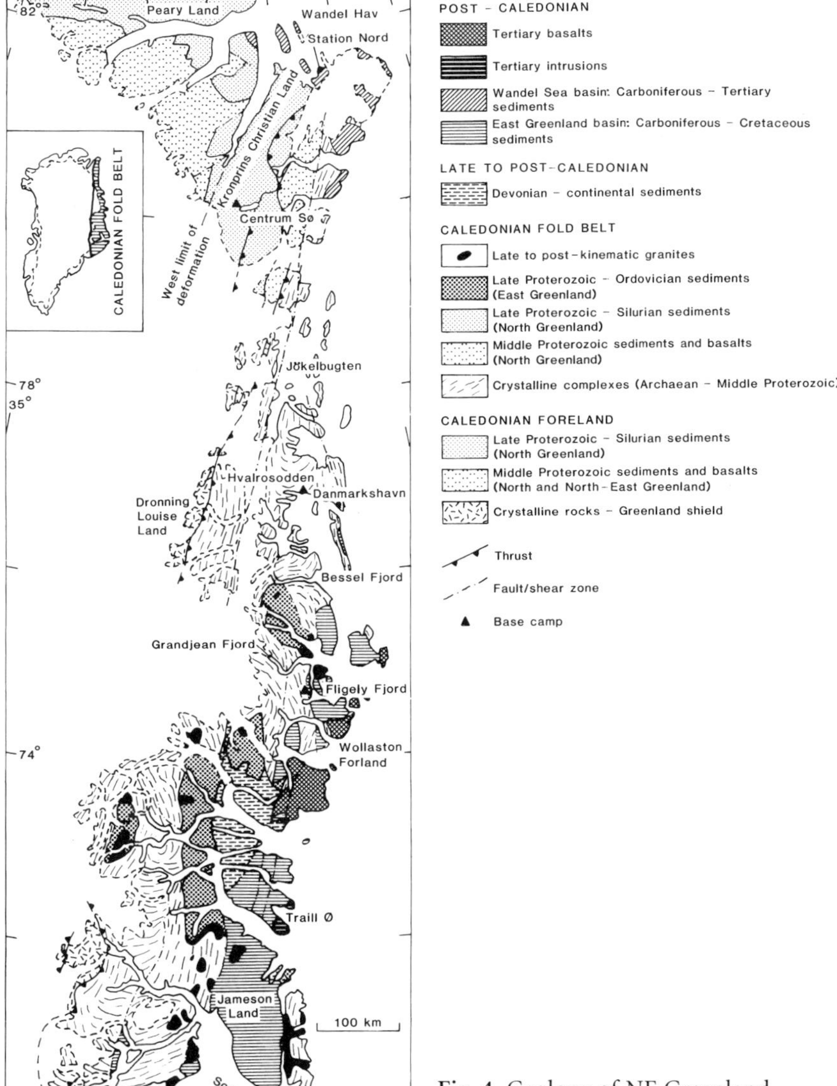

Fig. 4. Geology of NE.Greenland, from N. Henriksen (1994).

1961-1974	Danmarkshavn 76°46'N, 18°40'W		Daneborg 74°18'N, 20°13'W		Mestersvig 72°15'N, 23°54'W	
	Range	Av.	Range	Av.	Range	Av.
Mean temperature, °C, year	-11.1 - -13.1	-12.4	-6.5 - -12.0	-10.3	-9.3 - -11.6	-10.2
Mean temperature, June	-0.3 - +2.9	1.0	0.2 - 2.6	1.3	1.1 - 4.4	2.2
Mean temperature, July	2.7 - 5.0	3.7	2.6 - 4.9	3.8	3.7 - 7.3	5.6
Mean temperature, August	1.2 - 3.7	2.3	2.4 - 4.7	3.4	4.1 - 6.8	5.4
Yearly precipitation, mm	63.4 - 210.9	123	122.8 - 304.6	205	129.2 - 497.2	288

Table 1. Temperature and precipitation at three E.Greenland meteorological stations. Source: The yearly "Provisional mean temperatures and total amount of precipitation in mm. Greenland" edited by the Danish Meteorological Institute. Mean temperatures for July-August and the year missing for Mestersvig 1966.

podzol-like soils, meadow tundra soils, protoranker soils, tundra ranker soils, and soils associated with turf-hummocks and patterned ground features (Ugolini 1966 a, b). For the 73°N area, Sørensen (1935) discusses soil forming processes on patterned ground but no descriptions of soil profiles are presented other than sporadic remarks in his field notes. Generally, the intensity of horizon differentiating soil forming processes in well-drained soils increases from the outer coast to the middle fjord area, actually the Zackenberg valley, where podzols are widespread (Jakobsen 1992 b). There, the combination of a fairly long growing season and water sufficiently accessible for biological processes explains the more luxuriant plant cover, compared with both the outer coast and the warmer but arid inland (Jakobsen 1992 a). Preliminary investigations indicate a thickness of the active layer between 40 and 110 cm (Jakobsen 1992 a, b).

2.3. Climate

The climate is high arctic with mean yearly temperatures just below -10°C and mean temperatures above 0°C in only three months, as registered at the only three meteorological stations that have been working for longer periods (Table 1). Unfortunately, the marked increase in continentality towards the head of the fjords is not registered, as all three stations are situated close to the outer coast (Figs. 1-2). Here, dense fog from the sea often covers the lower 100-300 m of the land.

Sørensen (1933) mentions that the greatest luxuriance in vegetation is encountered from 300-600 m a.s.l. Thus, on the outer parts of Traill Ø easterly winds, often with fog, are dominating at sea level, whereas warm, often foehn-like, westerly winds dominate at 400-500 m a.s.l. The effect of fog was also clearly demonstrated in the observations at Nørrefjord (71°N) north of Scoresby Sund where, at the sea level, Salix arctica was in leafing or in the very beginning of flowering but only 150-200 m further upslope it was spreading its seeds (Fredskild & al. 1986).

A hint of the increasing continentality inland may be given by the meteorological observations made 1932-34 on Eskimonæs close to the outer coast, and on Ella Ø midway in the fjord system (Fig. 2). The winter was slightly colder on Ella Ø, but here the mean temperatures for June (3 years observations) were 2.6°C higher, July 4.2°, August 5.7°, and September 4.5°C higher (all 2 years obs., Sørensen 1941).

As clearly reflected in the position of the snowdrifts, the prevailing winds are northerly everywhere in East and Northeast Greenland. In 1951-1960 winds from NE-NNE-N-NNW-NW were registered at Daneborg in 40.5%, and calm weather in 30.7 % of the observations. At Danmarkshavn corresponding percentages were 39.2 and 27.7 (Lysgaard 1969).

2.4. Botanical exploration

Among the many botanical results of the "Three Year Expedition 1931-34 to Northeast Greenland", three papers deal with the taxonomy and distribution, geographically as well as phytosociologically, of the phanerogams from different areas: Sørensen (1933): 71°-73°30'N, Gelting (1934): 73°15'-76°20'N, and Seidenfaden & Sørensen (1937): 74°30'-79°N. These papers also include the material collected on earlier expeditions. Sørensen, in Seidenfaden & Sørensen (1937) presents a survey of the 18 main vegetation ecosystems and vegetation types in the area 71°-79°N, and for 130 of the most common species their occurrence in vegetation types are given. Gelting (1937) measured the snow cover regularly on four associations on Clavering Ø through 12 months from August 1931, and gives the distribution of a series of communities in relation to the snow cover in two transects. Sørensen (1945) summarises the botanical investigations in Northeast Greenland obtained during the Lauge Koch expeditions 1926-39.

After World War II, Schwarzenbach collected plants i.a. on many nunataks during short helicopter stops. In some papers Schwarzenbach (1951, 1960, 1961) gives a summary description of the plant communities of the inland at 73°-74°30'N. Between 1956 and 1964 Raup (1965, 1969 a, b) made thorough studies on the relation of flora and vegetation to moisture and physical disturbance gradients at Mestersvig (71°15'N). Elkington (1965) gives analyses from 15 Dryas dominated vegetations at Mestersvig.

In connection with oil exploration on Jameson Land, Greenland Botanical Survey conducted vegetation mapping for Greenland Environmental Research Institute in 1982-86. Based on many analyses, each consisting of 15 or 4 one m^2 spots with degree of cover for each phanerogam species following the Hult-Sernander scale, the major part of Jameson Land was subsequently mapped by false colour infrared aerial photos. Fourteen characteristic plant communities were recognised (Bay & Holt 1986), and the snowdepth was measured on 30 sites in March-April 1984 (Holt 1985). Fourteen analyses of Betula nana dominated stands are given in Fredskild (1991). Bay and Fredskild were the botanists on a three year project "Biological-Archaeological mapping of the Greenland east coast between 75° and 79°30'N" in 1988-1990, during which 55 localities were visited. Time only allowed for few vegetation analyses, but general descriptions are found in Bay & Fredskild (1990, 1991). In 1991-92 and 1996 the author worked in the Zackenberg area where a permanent field station has been established. 103 vegetation analyses are given in Fredskild & Bay (1993), 33 in Fredskild (1996). As to phanerogams, this area is the most diverse north of 74°N on both coasts of Greenland, with 150 species found so far.

Bay (1992) summarized all phanerogamic botanical expeditions to Northeast Greenland north of 74°.

3. Material and methods

3.1. Field methods

The vegetation was analyzed by Sørensen using the Raunkjær method, with twenty circles each of 0.1 m^2 , in the following termed a relevé. Besides, he scored which species were found within a central 0.01 m^2 circle. In 274 of the relevés 10 paper bags with cryptogams: mosses and/or lichens, and in some cases *Nostoc*, were collected.

In 1992 a 6640 m long line from the fjord to 610 m a.s.l. was laid out on the south side of Aucellabjerg in the Zackenberg valley by the author and G.S. Mogensen. The transitions between the different vegetations were marked with 106 permanent pegs, and a summary description made. Here, 63 relevés, consisting of 10 Raunkjær circles with regular intervals in homogenous vegetations along the line, were analyzed, with each circle consisting of three concentric circles: 0.1, 0.01, and 0.001 m^2. A species gets a score of 1, 2 or 3 points according to which circle it is registered, the smallest circle giving 3. In the tables in Fredskild & Bay (1993) and Fredskild (1996) besides the F% the total score for each species in the relevé is given, which informs on the species density. Elsewhere in the area 40 relevés were made by 10 random circles in homogenous communities. With one exception the analyses were made east of the river. In 1994 G.S. Mogensen extended the permanent line to the top of Aucellabjerg at 1040 m, and permanent plots with detailed registration of mosses and lichens have been established by Mogensen and E.S. Hansen (Fredskild & al. 1995). In 1996, 33 relevés were made, mainly west of the river (Fredskild 1996).

Using the Hult-Sernander scale, analyses were made in 1990 at Clausen Fjord (77½°N) at the head of Skærfjord in connection with ground truthing of satellite imageries showing the amount of biomass by different colours (Bay and Fredskild 1991). One of these analyses is included in Table 18.

3.2. Vegetation classification

Sørensen (1948: 33) describes "a method of grouping vegetation analyses on a basis of similarity of species content without any kind of subjective distinction between indicative species and companions, in other words: all species are being considered equally". The similarity between two relevés is expressed by the similarity quotient QS= 2c x 100/(a + b), in which a and b are the number of species in each of the relevés, c the number of species common to these. The results of Sørensen´s numerous calculations on the 392 relevés (of which he excluded 45) is visualized in a dendrogram, redrawn after his original (Fig. 3). The relevés were arranged in 41 groups, in which the internal QS exceeds 60. These 41 groups were arranged in wider groups: a-y with QS>48, A-N with QS>40, and I-VI with QS>30. Sørensen (s.a.) describes the latter six groups as: I herb-slopes, II snowbeds and permanently wet areas without closed vegetation, III fen and marsh, IV meagre dwarfshrub heaths on sandy soil, V mesic-dry dwarfshrub heaths, grass heaths and rock communities, VI clay, mainly marine, without closed vegetation. A-N which according to him (Sørensen s.a.) are comparable with alliances, were mostly referred to corresponding Scandinavian units (Nordhagen 1943). Sørensens designations for groups A-N and 1-41 have been retained in the present paper and the order mostly followed. The exceptions from this stems from the similarity index which does not consider the value of species being characteristic of or even selective of certain well defined plant communities. Sørensen also left a table, not presented here, of the similarity quotients between the 41 groups.

Daniëls (1994) gives a summary of the attempts in the past century to arrange the Greenland vegetation types into a hierachical system of the Braun-Blanquet method. The syntaxonomical classification and naming of the vegetation types in the present paper follow suggestions of Daniëls (1996, 1997 pers. comm.). As far as possible the vegetation units are classified in existing syntaxa of the Braun-Blanquet system (cf. Daniëls 1982, 1994, Dierssen 1996). The description of many new units will only be possible after ample comparison with, and evaluation of, many more vegetation analyses from other parts of Greenland and arctic regions. However, a few vegetation-types are described as new associations. The synoptical table (Table 36) gives a survey of the vegetation types. Here, the presence values r, +, and I-V (cfr. Daniëls 1982) is expressed in arabic figures: 1 (r, +, and I), 2 (II), 3 (III), 4 (IV), and 5 (V). The code of phytosociological nomenclature (Barkman & al. 1986) is principally followed.

As generally the number of relevés with cryptogam analyses are fewer, only the numbers 1-3 are used for presence values in Table 37 which only includes vegetation units with at least three relevés. In units with three relevés 1 means occurrence in one, 2 in two or three relevés. In units with four relevés 1 means occurrence in one or two, 2 in three or four relevés. In units with five or more relevés 1 means occurrence in less than one third of the relevés, 2 in one third to less than three fourth of the relevés, 3 in three fourth or more of the relevés. Species occurring in only one relevé in only one unit are excluded from the table. The species are arranged alphabetically, but the order of the vegetation units is the same as in the synoptical table.

If nothing else is mentioned, quotation marks in the text are used for translated quotations of Sørensen (s.a.) or his field notebooks.

3.3. Tables

Generally, species occurring in all units in a table are placed topmost, thus arranged after decreasing number of relevés in which they occur, and decreasing sum of frequency percentages (F%), followed by those occurring only in the first phytosociological unit, then in the first two, etc. The type relevé, or, in the case of variants, subtypes or communities, a typical relevé, is placed in the left column in a unit. This relevé is included in the following column which gives the average F%, and in the third, which gives the number of relevés (in italics) in which the species in question occur. Mostly, species occurring in only one relevé in a whole table are excluded but mentioned in the text under the unit in question, for type relevés and other relevés presented in full in the tables with the F% added. In some small tables all species are included.

In the Zackenberg relevés a species clearly belonging to a vegetation but seen only just outside the circles, is marked with a + in the tables. Such species are included in the number of species. In the cryptogam tables the number of species and sum of F% include taxa only determined to the genus level.

In addition to the number of species, the range and average sum of F% is given below in the tables. Obviously, the sum of F% per se does not inform on the degree of cover of a certain vegetation. However, the average and the range, of as well the sum of F% as the number of species, give an impression of the density and to a certain extent of the cover of the vegetation in question.

3.4. Taxonomical remarks

Generally, Böcher & al. (1978) is followed for phanerogams.

Cerastium arcticum
In the Scoresby Sund area besides *C. arcticum* also *C. alpinum* and intermediates with *C. arcticum* appear. Northwards to c. 74°N *C. arcticum* is all dominating, but the pubescence on the leaves of a few collections of *C. arcticum* may indicate hybrids/intermediates with *C. alpinum*. In accordance with Bay (1992) all material is termed *C. arcticum*.
Draba arctica
Seidenfaden and Sørensen (1937) did not

distinguish between *D. arctica* and *D. cinerea*, claiming that the species which they term *D. cinerea*, is a northern species. At Zackenberg, only *D. arctica* was collected, and Bay (1992) consider all Greenland material north of 74°N to be *D. arctica*. According to Böcher (1966), both *D. arctica* and *D. cinerea* occur in the area 70°-74°, with the former the most frequent. As some *D. cinerea* may be included in the analyses in the 73°N area, the term *Draba arctica/cinerea* is used in the tables for Sørensens material.

Draba adamsii

Sørensen (1933) and Seidenfaden and Sørensen (1937) discuss the *Draba micropetala-oblongata* complex and the difficulties in separating them. In the summary table, Sørensen uses only the term *D. micropetala*. In Böcher et al. (1978) both taxa are included in *D. adamsii*.

Dryas octopetala

Sørensen (1933) included var. *integrifolia* in *D. octopetala* and consequently all material in his lists is termed *D. octopetala*. According to the maps in Bay (1992), *D. octopetala* is very common and *D. integrifolia* very rare between 74°N and 78°N, but hybrids are frequent, confirmed by collections from Zackenberg. Elkington (1965) shows the highly variable Greenland material of *Dryas* as a result of an introgressive hybridization between *D. octopetala* and *D. integrifolia*. However, according to his maps (l.c. Figs. 17 and 18) *D. octopetala* is (all) dominating from 69°-78°N in East Greenland. In the tables the term *D. octopetala* is retained for the material from the 73°N area, whereas *Dryas* sp. is used for the Zackenberg tables because of the frequent hybrids there.

Festuca brachyphylla s.l.

Sørensen (1933) and Sørensen and Seidenfaden (1937) consider *F. ovina* identical with *F. brachyphylla* but realize that several forms occur. However, in his analyses Sørensen does not distinguish lower taxa of *F. brachyphylla*. As discussed below under Ass. Koenigio-Saginetum intermediae, *F. brachyphylla* in some communities undoubtedly are *F. hyperborea*. Besides, many of Sørensens collections from a.o. Ella Ø, Ymer Ø and Hold with Hope, by him labelled *F. brachyphylla*, are *F. baffinensis*. Because of this, the designation *F. brachyphylla* s.l. has been used in the tables for Sørensens material except in the table of Koenigio-Saginetum intermediae. At Zackenberg the three species have been separated. If a table includes analyses from both Zackenberg and the 73°N area, *Festuca brachyphylla/s.l.* means that the Zackenberg material is only *F. brachyphylla*, whereas other species may be included in the 73°N material. According to Aiken & al. (1995), *F. edlundiae*, closely related to *F. hyperborea*, occurs in Northeast Greenland. Thus, a specimen from a "denuded bog (microjunceon)" on Hold with Hope, coll. by Sørensen Aug. 14, 1934, and by him labelled *F. brachyphylla* var. ?, and later determined by K. Holmen as *F. hyperborea*, has been redetermined by Aiken to *F. edlundiae*.

Melandrium triflorum/affine

In Sørensens summary table is written: "*Melandrium affine* (incl. *triflorum*)". As the field determinations of the two species often causes difficulties and, besides, both species are equally frequent northwards to c. 78°N, *M. triflorum/affine* is used in the tables. In Table 29 all *Melandrium* in seven Zackenberg relevés (Z14, 15) are *M. triflorum*. Most likely only this species was represented in the Sørensen relevé of this association, too.

Poa arctica/pratensis.

In the analyses, Sørensen distinguish *Poa arctica* from *Poa* (*pratensis* ssp.) *alpigena*. This has been retained in the tables.

Potentilla hookeriana/nivea

In the 70°-76°N area the distribution of the northern *P. hookeriana* overlap that of the southern *P. nivea*, with which it often hybridizes (Bay 1992, maps 50 and 95). Therefore, the designation *P. hookeriana/ nivea* is used for the Sørensen relevés in the tables.

In all analyses from Zackenberg it is represented by *P. hookeriana*, *P. nivea* only seen once on a south facing, gneissic rock.

Saxifraga nivalis and *S. tenuis*
In the summary table, Sørensen separates *S. nivalis* and *S. nivalis* ssp. *tenuis*. In accordance with this, the designation in the tables is either *S. nivalis* or *S. tenuis*.

Saxifraga hyperborea/rivularis
Sørensen only uses the term *S. rivularis*. However, in Sørensen (1933) he mentions a "slender form, shoot with red and with small reddish corollas.....analogous to the var. *tenuis* Wahlenb. of *S. nivalis*, in company with which plant it often occurs". In 1953, he redetermined a number of his 1933-34 collections in Herb.C, labelled *S. rivularis*, to *S. hyperborea*. *S. rivularis* becomes rare towards its N-limit at 78°N, whereas the circumgreenlandic *S. hyperborea* is very frequent between Scoresby Sund and 78°N. In the tables the term *S. hyperborea/rivularis* is used.

Stellaria longipes
In accordance with Bay (1992) *S. longipes*, which includes several lower taxa, is treated as one taxon, in the tables designated *S. longipes* s.l.

Abbreviations used in the tables:
Carex pseudolagopina for C. marina ssp. *pseudolagopina*, *Empetrum hermaphroditum* for *E. nigrum* ssp. *hermaphroditum*, *Vaccinium microphyllum* for *V. uliginosum* ssp. *microphyllum*, and *Polytrichum strictum* for *P. juniperinum var. strictum*.

The taxonomy for mosses used by K. Holmen have been updated by G. S. Mogensen. For lichens the taxonomy follows Hansen (1995). All collections of *Solorina* were by Holmen termed *S. octospora*. According to E.S. Hansen, who determined the lichens from the Zackenberg analyses, the material collected by Sørensen may well include other species, especially *S. saccata* and *S. bispora*. As to *Stereocaulon*, Holmen determined *S. alpinum* and *S. paschale*, but those collections determined as the latter presumably include other species, e.g. *S. rivulorum* and *S. groenlandicum* (E.S. Hansen, verbal information 1996, see also Hansen 1996).

3.5. Soil analyses

In addition to the vegetation analyses, Sørensen collected soil samples. In 1960-61 the Danish National Science Research Foundation granted 2 years technical assistance to M. Køie and Th. Sørensen to carry out analyses of Greenland soil samples, half of which were from Northeast Greenland. In 1963 a further grant was given for computer treatment of the punched cards with results of the many analyses that had been undertaken at the Department of Plant Ecology. A total of 268 of Sørensens topsoil samples were analyzed. The methods were:

Specific conductivity: 10 g soil + 50 ml water, expressed in µ mho at 20°C. Ca, Mg, K, Na, Mn: Exchangeable contents, extracted by means of 1 N NH_4OAc. K and Na measured by a Coleman Flame Photometer 21. Ca, Mg and Mn measured by a Perkin-Elmer Atomic Absorption Spectrophotometer 303. Expressed in meq/100 g dry soil.

P: Extracted by means of 0.2 N H_2SO_4. Measured according to the ammonium molybdate method by a Coleman Spectrophotometer Universal 14 (Jackson 1958). Expressed in ppm dry soil.

Cu, Ni, Zn, Pb: Extracted by means of 0.02 M EDTA, 2Na (Titriplex III). 20 g soil + 50 ml EDTA. Measured by a Perkin-Elmer Atomic Absorption Spectrophotometer 303. Expressed in ppm dry soil.

pH: 10 g soil + 50 ml water. - In 1931-33 Sørensen measured pH on some samples, mostly of the topsoil, sometimes also of the subsoil on the Ella Ø station. These results, given in his list of vegetation analyses, are referred to under the description of the plant communities.

Organic matter: Wet combustion of communinuted soil (Walkley-Black's method, cf. Jackson 1958). Expressed in percentages by weight of soil.

Li, Sr, Fe and cation capacity, not considered here, were measured on some samples.

3.6. Plant to soil relationships

Calculation of the correlations between each plant species and each soil factor was carried out by means of a computer programme elaborated by M. Køie. This programme calculates, for each species, the percentage distribution of observations (sample plots with each species) at four levels of each soil factor. Each level represents an interval in the variation of the particular factor. Limits between the levels are fixed in such a way that an equal number of data occur within each of the four categories. Accordingly, levels obtained are not always equidistant. The method is further discussed in connection with a statistical treatment of 19 soil factors from 483 danish sample plots in Hansen & Jensen (1974). This paper also demonstrates significant linear correlations by regression equations, corresponding correlation coefficients, and significance levels for the combinations of edaphic factors.

In connection with the planned publication in 1976-78, the most important pedological factors for 109 phanerogam species were visualised by drawing figures of the type presented in Hansen & Jensen (1974). 40 illustrations showed nine factors for each species, and 69 showed only pH and conductivity. With the exception of the three most ubiquitous species (*Dryas octopetala, Polygonum viviparum,* and *Salix arctica*), these 69 species are mostly occurring in quite few relevés. Nine factors are shown only for species with at least 24 sample spots. The four levels chosen for each soil factor are given (Fig. 5). On top of the frame the species name and the number of analyzed sample plots in which it occurs is indicated, and on the left side the percentages. For each soil factor the mean value is indicated. A statistical treatment of the correlation between the edaphic factors was not finished. However, some correlations are clearly read from the figures, e.g. the positive correlation between pH, conductivity and Ca, and the high values of Cu being sometimes correlated with low pH (*Hierochloë alpina* and *Poa arctica*), sometimes with high pH (*Minuartia stricta* and *Braya purpurascens*). The range of the different factors obtained in the samples from the 73°N area (Table 2) are comparable with those few papers published on Greenland soils informing on more than pH and conductivity (Hansen 1969, Tedrow 1970, Holowaychuch & Everett 1972, Stäblein 1977, Jakobsen 1988). As changes throughout the soil profiles are not elucidated in the samples from the 73°N area, only the generel impression of the different soils on which the plant communities are found will be mentioned, based on the Sørensen field notes and on the ecological preference of those species characterizing the community in question. If a species is found in at least 10 sample plots the figure is shown with its edaphical curves. The curves for the most frequent, ubiquitous species are given in Figs 6-7, clearly illustrating their independence of most factors by having c. 25% of their occurrence at each of the four levels.

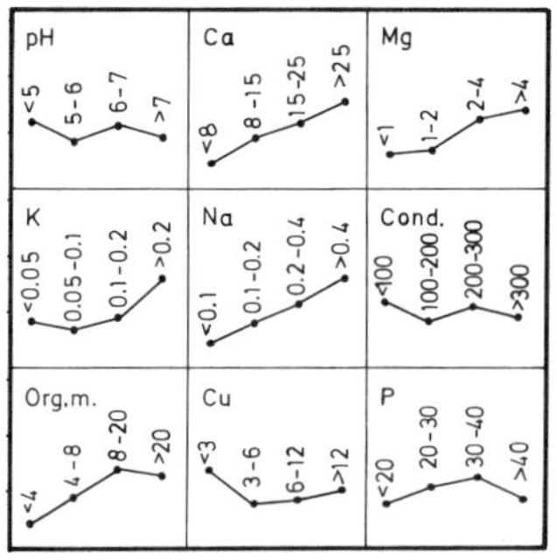

Fig. 5. The four levels used for each of nine soil factors, shown for different plant species in Figs. 6-26.

	Minimum	Maximum
pH	*Hierochloë alpina*: 4.88 (24) *Poa arctica*: 5.25 (54)	*Lesquerella arctica*: 7.02 (21) *Braya purpurascens*: 6.92 (28)
cond.	*Saxifraga tenuis*: 69 (30) *Sagina intermedia*: 73 (23)	*Carex saxatilis*: 480 (21) *Saxifraga nathorstii*: 441 (27)
Ca	*Saxifraga tenuis*: 8.45 (30) *Hierochloë alpina*: 9.73 (24)	*Carex parallela*: 25.9 (35) *Kobresia simpliciuscula*: 25.0 (52)
Mg	*Minuartia biflora*: 1.73 (35) *Trisetum spicatum*: 1.81 (33)	*Carex parallela*: 3.64: (35) *Kobresia simpliciuscula*: 3.53 (52)
K	*Saxifraga tenuis*: 0.079 (30) *Braya purpurascens*: 0.089 (28)	*Hierochloë alpina*: 0.201 (24) *Pyrola grandiflora*: 0.170 (33)
Na	*Saxifraga tenuis*: 0.149 (30) *Minuartia rubella*: 0.162 (38)	*Carex parallela*: 1.27 (35) *Saxifraga aizoides*: 1.23 (50)
Organic matter	*Saxifraga tenuis*: 3.28 (30) *Minuartia biflora*: 3.98 (35)	*Eriophorum triste*: 14.2 (65) *Pyrola grandiflora*: 14.0 (33)
Cu	*Minuartia stricta*: 4.58 (29) *Minuartia rubella*: 5.47 (38)	*Hierochloë alpina*: 14.1 (24) *Eriophorum triste*: 12.8 (65)
P	*Carex bigelowii*: 28.6 (76) *Carex capillaris*: 29.1 (60)	*Saxifraga tenuis*: 39.0 (30) *Minuartia biflora*: 38.8 (35)

Table 2. Range of average values of nine pedological factors. Numbers in brackets give the number of sample spots. For each factor the two lowest/highest values are given. Only species with at least 20 sample spots are considered.

3.7. Species to species relationships

Supplementing the plant-soil relations, lists of species to species (phanerogam to phanerogam, and moss to moss) relations were made. For each species occurring in at least 18 relevés, a list was made of the 15 species with which it most often grows together. For example, the relation *Betula nana-Pedicularis lapponica* is 100%, as *Betula nana* is found in each of the 40 relevés with the latter. And in 49 of 52 relevés with *Kobresia simpliciuscula* there is also *Carex misandra*. However, for many species too few data are available to allow for a statistical treatment, and no tables are given.

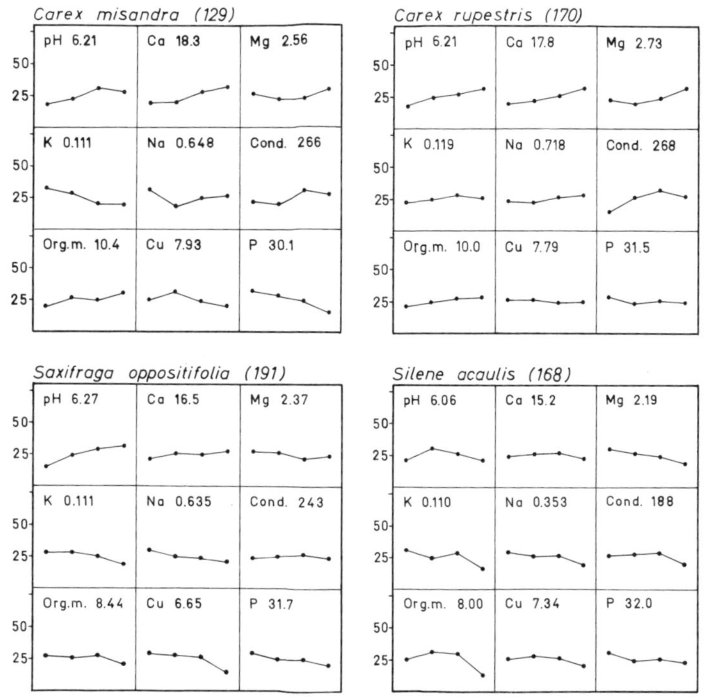

Fig. 6. Soil characteristica for four common, ubiquitous species.

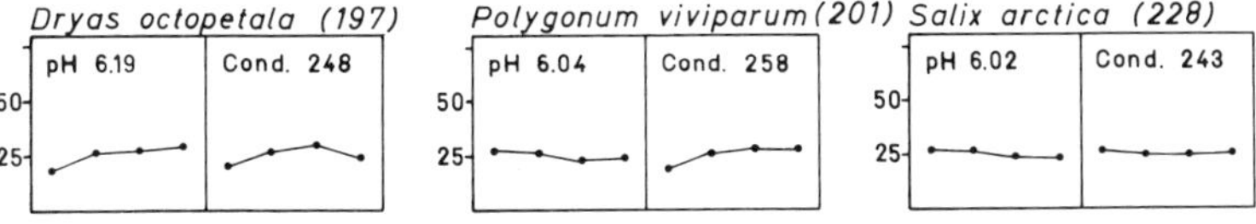

Fig. 7. pH and conductivity for three common, ubiquitous species.

4. All. Saxifrago -Ranunculion nivalis Nordh. 1943 emend. Dierssen 1984

4.1. Middle-arctic, meso - hygrophilous herb-slope vegetation on early snowbeds

As a result of the prevailing northerly winds during winter the snow accumulates on south facing slopes. Locally, a belt just below the top of the ridges but above the late snowbeds, may harbour a comparatively lush vegetation rich in forbs, if the soil remains moist throughout the summer.

Sørensen (s.a.) groupes 15 relevés from the 73°N area under (translated) "Alliance A representing the herb-slope vegetations which, if anything, correspond to Nordhagens Junceon trifidi scandinavicum, yet with affinity to Nardeto-Caricion rigidae. Typically, these herb-slopes are only developed in the Middle Fjord zone (a); towards the outer coast they are atypical (b) with a close affinity to snowbed vegetations. Totally missing in the Inner Fjord Zone". However, since both alliances mentioned are found on definitely acid ground, and graminoids are dominating, it is included in suball. Luzulenion arcticae (Nordh. 1936) Gjærevoll 1956. It is characteristic for the Middle Arctic Tundra Zone (sensu Elvebakk 1985) of Northeast Greenland. Corresponding thermophilous *Trisetum* snowbeds on Spitsbergen, the ass. Oxyrio-Trisetetum spicati, are included in Ranunculo-Oxyrion by Hadac (1989).

With "Alliance A" included in Saxifrago-Ranunculion, the characteristic species for this alliance in Northeast Greenland are: *Trisetum spicatum, Oxyria digyna, Minuartia biflora, Ranunculus pygmaeus, Cerastium arcticum*, and *Ranunculus sulphureus*. Differential species for suball. Luzulenion arcticae against suball. Drabo-Cardaminenion bellidifoliae are: *Antennaria canescens, A. porsildii, Erigeron humilis, Festuca rubra, Luzula spicata, Poa alpina, Potentilla crantzii*, and *Taraxacum brachyceras*.

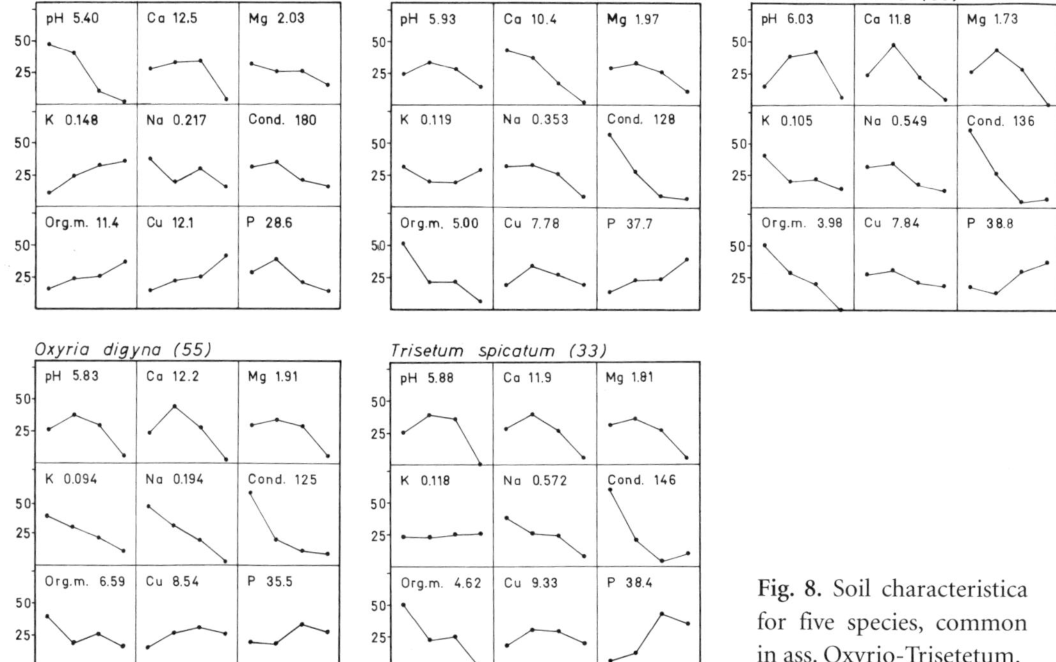

Fig. 8. Soil characteristica for five species, common in ass. Oxyrio-Trisetetum.

Floristically, the alliance differs fairly much from other alliances. Thus, in Sørensens analyses the following species are only recorded here: *Antennaria canescens, A. porsildii, Arabis alpina, Botrychium lunaria, Gentiana nivalis, G. tenella, Harrimanella hypnoides, Luzula spicata, Poa alpina* var. *vivipara, Potentilla crantzii, Sibbaldia procumbens, Taraxacum brachyceras, Tortula norvegica*, and in only one relevé outside: *Draba crassifolia*.

The suball. is found in the area between Scoresby Sund and ca. 75° N, illustrated by the many low- or middle arctic species having their N-limit af 74°-76°, i.e. the phytogeographical distribution type 14 of Bay (1992). Judging from the ecological preferences of the character and differential species (Figs 8-9) pH is mostly around 6, conductivity and content of organic matter low.

Ass. Oxyrio-Trisetetum Hadac (1946) 1989. (Z1, A1-2)

Characteristic species for the association (Table 3) are *Trisetum spicatum, Minuartia biflora*, and *Erigeron humilis*. Southern species like *Potentilla crantzii, Poa alpina, Antennaria canescens, Taraxacum brachyceras, Festuca rubra*, and *Euphrasia frigida* occur in some units. Generally, mosses are few.

Variant of *Saxifraga oppositifolia* (A1)

Five relevés from middle altitude, south facing slopes in the middle Fjord Zone are from fairly dry herb-rich vegetations, as indicated by the differential species *Saxifraga oppositifolia* and *Dryas* sp. in all, and *Carex nardina* in three relevés. The soil is neutral (pH 6.4-7.1, 5 anal.), not totally

Association	A										
Variant/community	A1			A2			Z1			A3	
Area	EY			T			Z			HH	
Elevation, m	275-600			175-475			30-300			0-100	
No. of relevés			*5*			*8*			*6*		
F% (F) or average F% (av.)	F	av.		F	av.		F	av.		F	F
Analysis no.	329			231			65			44	46
Trisetum spicatum	100	80	*5*	90	77	*8*	80	73	*6*	100	20
Polygonum viviparum	85	66	*5*	70	77	*8*	60	60	*6*	100	100
Salix arctica	60	67	*5*	20	58	*8*	50	72	*6*	10	100
Minuartia biflora	95	45	*4*	100	81	*8*	90	60	*6*	100	60
Silene acaulis	40	53	*5*	70	58	*7*	40	35	*5*		10
Erigeron humilis		31	*3*	25	44	*8*	90	48	*4*	100	40
Saxifraga cernua		44	*4*	25	19	*7*	80	30	*4*	10	100
Oxyria digyna	100	56	*4*	100	64	*7*	60	45	*4*	100	80
Ranunculus pygmaeus		14	*1*	90	49	*7*	50	28	*4*	100	100
Equisetum arvense		8	*1*	80	41	*5*		7	*2*	70	
Saxifraga oppositifolia	20	51	*5*		1	*1*		+	*1*		40
Minuartia rubella		18	*2*								
Minuartia stricta		2	*2*								
Potentilla crantzii	95	22	*2*	100	73	*8*					
Poa alpina	100	41	*3*	100	51	*7*					
Antennaria canescens	65	45	*4*	65	30	*5*					
Carex scirpoidea		29	*2*	20	18	*6*					
Taraxacum brachyceras	25	8	*3*	75	35	*4*					
Antennaria porsildii		19	*1*	20	29	*5*					
Euphrasia frigida	25	6	*2*		19	*2*					
Pedicularis flammea		2	*1*		1	*1*					
Cerastium arcticum	80	31	*5*	95	61	*7*	80	58	*6*		
Festuca rubra	100	59	*4*	95	83	*7*	100	22	*3*		
Draba glabella	100	69	*5*	60	8	*2*	20	10	*4*		
Equisetum variegatum		10	*1*		17	*4*		2	*1*		
Arnica angustifolia		9	*1*		2	*1*	30	5	*1*		
Dryas octopetala/sp.	25	52	*5*					10	*3*		
Carex nardina	20	16	*3*					+	*1*		
Melandrium triflorum/affine		7	*2*				+		*1*		
Carex rupestris		14	*1*					5	*1*		
Poa glauca		2	*2*					2	*1*		
Papaver radicatum		1	*1*					+	*1*		
Luzula spicata				55	47	*6*					
Thalictrum alpinum					38	*5*					
Carex lachenalii				80	39	*4*					
Poa pratensis ssp. *alpigena*				25	18	*4*					
Sibbaldia procumbens				90	33	*3*					
Salix herbacea					25	*2*					
Gentiana tenella					14	*2*					
Draba fladnizensis					13	*2*					
Draba crassifolia				30	5	*2*					
Juncus biglumis					4	*2*					
Carex bigelowii				100	99	*8*		30	*2*		
Poa arctica					12	*1*	60	57	*6*		
Draba lactea					3	*2*	20	10	*4*		
Ranunculus sulphureus					3	*2*	20	13	*2*		
Saxifraga nivalis					2	*1*		2	*1*		
Draba alpina					4	*2*					40
Potentilla hyparctica							50	45	*5*		
Luzula confusa							50	30	*4*		
Alopecurus alpinus							40	18	*4*		
Stellaria longipes s.l.							30	7	*2*		
Festuca brachyphylla								7	*2*		
Hierochloë alpina								3	*2*		
Taraxacum arcticum								2	*2*		
Pedicularis hirsuta								+	*2*		
Phippsia algida										30	70
Sagina intermedia											100
Cerastium regelii											90
Poa alpina var. *vivipara*											30
No. of species, range	17-24			16-27			15-23			11	15
No. of species, average	20			23			20				
Summa F%, range	815-1205			1045-1750			550-1120			730	980
Summa F%, average	995			1376			818				

Table 3. Ass. Oxyrio-Trisetetum, phanerogams.

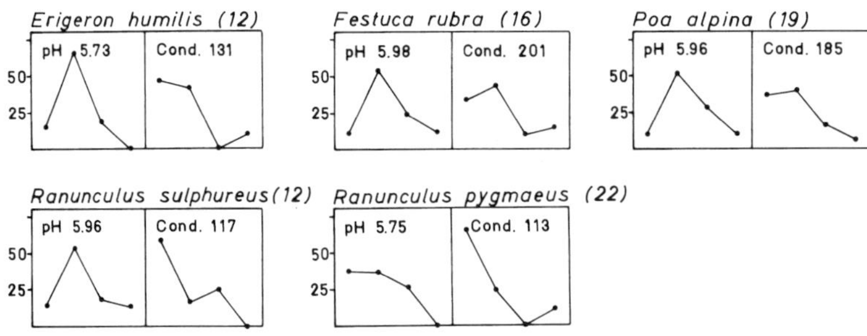

Fig. 9. pH and conductivity for five species, common in ass. Oxyrio-Trisetetum.

covered by vegetation as indicated by the average Sum of F%: 995 (phanerogams) and 263 (cryptogams). A typical analysis (Tables 3-4, anal. 329), is from a slope 500 m a.s.l. A closed, lichen rich *Dryas* heath with i.a. *Antennaria porsildii* and *Arnica angustifolia*, 30 km southeast of Mestersvig is representative of the variant (Elkington 1965, anal. 2).

In one relevé only: *Braya humilis* (10%) and *Lesquerella arctica* (25%), both from the typical relevé, and *Arenaria pseudofrigida, Betula nana, Carex capillaris, Saxifraga aizoides.*

Variant of *Potentilla crantzii* (A2)

Beyond doubt a result of the fairly continental climate of Traill Ø the herb-slopes here are most luxuriant, shown by large diversity and the highest but one average sum of F% (1376) for phanerogams of the 41 groups from the 73°N area. All relevés are from lower elevations, the five 175-200 m. The soil is acid (pH 5.0-5.6, 8 anal.), as also indicated by the occurrence of *Carex bigelowii* (Fig. 8). A typical analysis (Tables 3-4, anal. 231) is from a south slope 200 m a.s.l. In the 73°N area *Tortula norvegica* was only collected in this variant (Table 4).

In one relevé only: *Arabis alpina, Botrychium lunaria, Cassiope tetragona, Gentiana nivalis, Harrimanella hypnoides, Saxifraga hieraciifolia.*

Variant of *Poa arctica* (Z1)

At Zackenberg the few, small and more poor herb-slope vegetations are all found at lower elevations, on the southern slopes of moraines, along brooks, or on the front of stabilized solifluction lobes. Usually, only a few m² are covered. Generally, they are downslope replaced by true snowbed vegetations. Among the species characterizing the herb-slopes in the 73°N area, having their N-limits between 74° and 76°N, the seven were not found at Zackenberg: *Antennaria alpina, A. porsildii, Luzula spicata, Potentilla crantzii, Taraxacum brachyceras, Poa pratensis ssp. alpigena*, and *Sibbaldia procumbens. Poa alpina* was only found once, growing with *Minuartia biflora* and *Ranunculus pygmaeus* on a SW-slope. *Festuca rubra* is here very rare, exclusively growing in herb-slope vegetations. On the contrary, the differential species, viz. the northern *Potentilla hyparctica* and the circumgreenlandic *Poa arctica* are frequent, growing in many plant communities. A

Association	A					
Variant	A1			A2		
No. of relevés			*4*			*3*
F% (F) or average F% (av.)	F	av.		F	av.	
Analysis no.	329			231		
Distichium capillaceum	30	53	*4*	30	33	*2*
Cladonia pyxidata		33	*2*	10	7	*2*
Bryoerythrophyllum recurvirostre	40	17	*3*			
Ditrichum flexicaule		30	*2*			
Nostoc sp.	20	8	*2*			
Stegonia latifolia		13	*1*			
Polytrichastrum alpinum				60	80	*3*
Tortula norvegica				20	13	*2*
No. of species, range	4-9			4-8		
No. of species, average	6			6		
Summa F%, range	180-320			120-300		
Summa F%, average	263			227		

Table 4. Ass. Oxyrio-Trisetetum, cryptogams

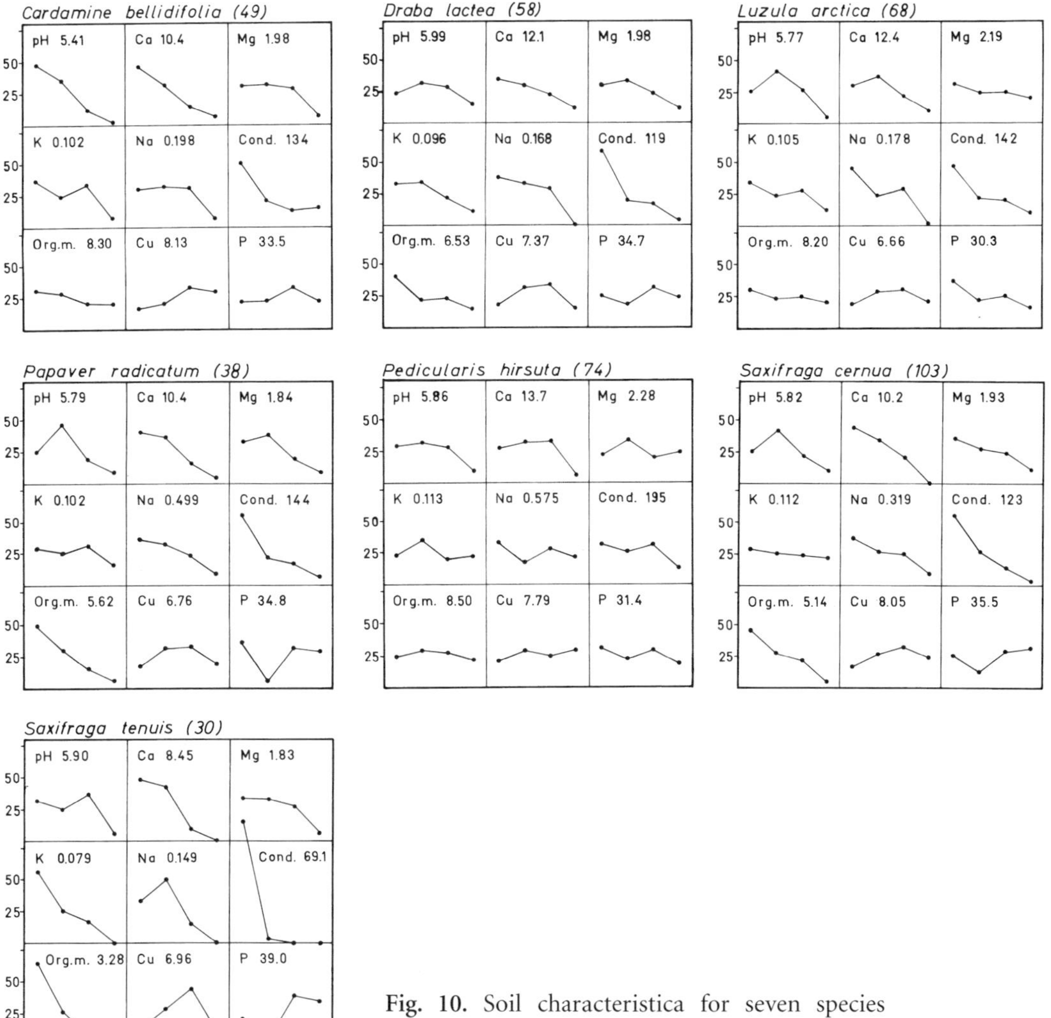

Fig. 10. Soil characteristica for seven species common in ass. Phippsietum algidae-concinnae.

typical analysis (Table 3, anal. 65) is from a 25-30° SW-slope on the front of a giant solifluction lobe, 300 m a.s.l. The six relevés are given in Fredskild & Bay (1993, Table 5, anal. 64-69). In one of these *Melandrium affine* (+) is found.

In one relevé only: *Melandrium apetalum* and *Erigeron eriocephalus,* both 10% in the typical relevé, and *Draba arctica, Festuca hyperborea, F. vivipara, Ranunculus nivalis, Saxifraga tenuis,* and *Vaccinium uliginosum ssp. microphyllum.*

Ranunculus pygmaeus-Trisetum spicatum community (A3)
No typical herb-slopes were found at the outer coast in the 73°N area. The two relevés from Hold with Hope (Table 3) were clearly from sites with longer snow cover, more humid soil and cooler conditions, thus resembling more northerly communities. The occurrence of *Phippsia algida, Erigeron humilis,* and *Trisetum spicatum* illustrate the intermediate position between snowbeds and herb-slopes, as stated by Sørensen (s.a.).

4.2. Middle-arctic snowbed vegetation

Sørensen (s.a.) groups 30 relevés under "Alliance B, including the fairly eutrophic snowbed vegetation corresponding to Nordhagens Ranunculeto-Oxyrion. It is well represented at the outer coast and

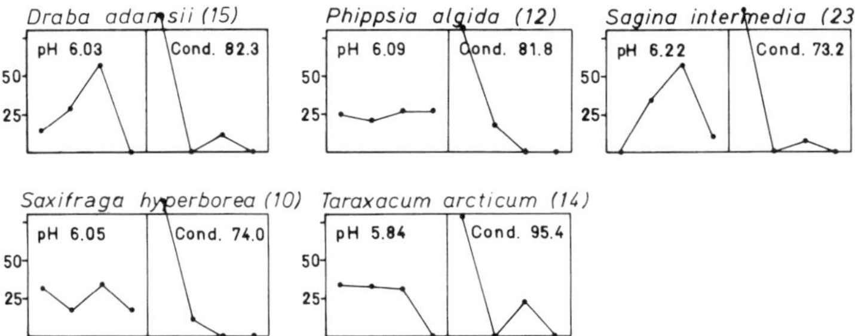

Fig. 11. pH and conductivity for five species common in ass. Phippsietum algidae-concinnae.

especially in the outer part of the middle fjord area. At the head of the fjords only fragmentarily at high altitudes. Characteristic species are *Draba micropetala, Saxifraga tenuis, Sagina intermedia, Taraxacum arcticum*". It is here divided into the late snowbed vegetations, including Phippsietum algidae-concinnae Nordh. 1943 subass. typicum and subass. oxyrietosum digynae, the latter with three variants and one community, and the moderate snowbed vegetations represented by Luzulo-Salicetum herbaceae with two variants. These associations are characteristic of the Middle Arctic Tundra Zone in Greenland. Differential species against the units of Oxyrio-Trisetetum are *Cardamine bellidifolia, Draba adamsii, D. lactea, Juncus biglumis, Luzula arctica, L. confusa, Minuartia rubella, Poa arctica, Phippsia algida, Sagina intermedia, Saxifraga tenuis, Taraxacum arcticum*, and, among cryptogams *Bartramia ityphylla, Cetrariella delisei*, and *Polytrichum juniperinum*. Apart from here, *Bartramia ityphylla* has only been found in one relevé in three associations in other alliances. Towards the northern limit of the zone fragmentary vegetations of the suballiance represent the northernmost early snowbeds. Thus, on Lambert Land (79°10′N) a species rich moss-snowbed with i.a. *Taraxacum arcticum, Papaver radicatum, Cerastium arcticum, Saxifraga tenuis, S. nivalis, S. cernua, Ranunculus pygmaeus, R. sulphureus, Melandrium apetalum, Draba lactea, Festuca hyperborea*, and *Trisetum spicatum* was found on a south facing slope 375 m a.s.l. And an early snowbed 400 m a.s.l. harboured i.a. *Oxyria digyna, Potentilla hyparctica, Taraxacum arcticum, Sagina intermedia*, and *Festuca hyperborea* (Bay & Fredskild 1991). Similar "high arctic herb-slopes" (Bay 1992) may locally cover a few square metres in the interior North Greenland.

On Svalbard the moderate snowbeds have a characteristic element of weakly thermophilous species such as *Trisetum spicatum, Minuartia biflora, Erigeron humilis, Ranunculus pygmaeus, R. nivalis*, and *Taraxacum arcticum*, and the substrate is mostly weakly acid (Elvebakk 1994). Quite many species of Oxyrio-Trisetetum spicati on Svalbard are in common with the Northeast Greenland snowbeds of the Alliance, whereas those from Southeast Greenland differs in their many, southern species (de Molenaar 1976).

Judging from the ecological preferences of most character and differential species (Figs 10-11) the soil is slightly acid, conductivity very low, a result of the seeping meltwater with an extremely low conductivity. *Cardamine bellidifolia*, occurring in 27 of the 48 relevés, has a marked lower pH preference.

4.2.1. Late snowbed vegetation

Ass. Phippsietum algidae-concinnae Nordh. 1943. (Z2, B4)

Characteristic species are *Saxifraga hyperborea/rivularis*, in the 73°N area only occurring in this subassociation, and *Phippsia algida*.

Subass. typicum subass. nov. (Z2)

Below the late melting snowdrifts at Zackenberg, often lying at the transition from a south facing slope to the almost

						B4						B5							
Subass./variant/community	Z2			B4.1	B4.2		B4.3			B5.1		B5.2		B5.3			Z3		
Area	Z			H	T		HH			T		T		HH			Z		
Elevation, m	2-90			275-600	330-580		<100			450-600		60-525		<100			5-310		
No. of relevés			7	3		7		4	14		5		7		4	16			11
F% (F) or average F% (av.)		av.		av.	F	av.	F	av.		F	av.	F	av.	F	av.		F	av.	
Analysis no.	106				144		16			161		199		43			56		
Luzula confusa	70	54	6	62	100	76	90	57	14	75	76	50	46	20	47	15	40	66	11
Salix arctica		9	3	97	95	75	90	35	13	55	79	100	100	100	90	16	90	95	11
Saxifraga tenuis	20	21	5	32	80	73	80	45	14	70	27	35	16	10	5	14	+	12	7
Saxifraga cernua		1	1	67	90	75	90	95	13	75	84	65	56	40	30	16	40	37	10
Sagina intermedia		9	2	2	50	56	50	62	11	30	25	25	15		10	14	20	18	6
Saxifraga oppositifolia		7	2	55		2		15	7	40	35	20	56	100	59	14		9	3
Ranunculus pygmaeus		1	1	52	30	35		38	8	100	45		1		5	5	50	31	7
Saxifraga hyperborea/rivularis	60	57	6	43	30	38		10	10						2	1		3	2
Cardamine bellidifolia		4	1	28	80	31		17	7		27	10	8	10	16	9		9	3
Stellaria longipes s.l.	10	3	2			16	40	37	6				1		21	4	20	18	8
Saxifraga foliolosa	20	29	3			13	30	20	3						4	1	20	15	3
Phippsia algida	100	91	7	62		39	60	67	11									6	4
Equisetum arvense		+	1												12	1		9	1
Alopecurus alpinus		1	2														60	45	11
Draba lactea				40	20	14		5	9	15	62	85	63	80	54	15	30	33	10
Oxyria digyna				27	100	79	70	47	12	100	58	30	29		7	11	10	15	8
Juncus biglumis				17		19	70	35	7		17	10	46	80	60	12	+	11	4
Festuca brachyphylla/s.l.				23	15	4	40	15	5		19		6		16	6		5	1
Trisetum spicatum				57	15	16	20	20	8	80	72	5	4		6	9	20	22	7
Taraxacum arcticum				3		11		4	3	70	57		1	50	20	7	10	6	6
Minuartia rubella				2	20	6		5	5		44	30	21	10	28	11		1	1
Cerastium arcticum				18	95	66	80	28	12	80	70	85	54	40	34	16	10	20	9
Silene acaulis				10		3			7		15	5	31	30	24	12	+	6	3
Poa arctica				32	20	8			6	90	32	5	5		4	6	50	58	11
Polygonum viviparum				8		2			4		20		29	100	76	7	90	25	6
Draba nivalis				3		1			2		8					1			
Papaver radicatum				2				8	2		31	5	11		15	11			
Potentilla hyparcticum				30				2	2		17			40	37	4	10	24	6
Draba adamsii					95	41	20	53	10	10	8	50	19		8	6			
Ranunculus sulphureus						1	20	5	2		29		1		1	4			
Luzula arctica					30	11	100	42	7	15	14	40	33	20	45	13	30	27	9
Draba alpina						14	70	23	4		20				5	4			
Cochlearia groenlandica						1	50	65	5										
Minuartia biflora						19			4	25	48	65	45		10	11		5	3
Carex misandra						1			1	5	17	10	53	90	26	12	10	2	3
Pedicularis hirsuta						1			2		5	55	76	40	21	12	20	4	3
Ranunculus nivalis						14			1						6	1	10	4	4
Erigeron humilis						1			1		18					1			
Cerastium regelii							90	60	3										
Draba subcapitata								2	1	95	26				9	4			
Poa glauca								5	1		2			10	2	2			
Carex nardina											1	5	9	20	16	11		1	1
Melandrium apetalum										20	7		10		1	7			
Draba glabella											19		5			4			
Festuca rubra											38					2			
Cassiope tetragona										5	2					2		1	1
Saxifraga caespitosa											1			10	2	2			
Dryas octopetala/sp.												5	4	20	9	8	20	14	3
Minuartia stricta													19			4			
Saxifraga hieraciifolia													6			2			
Festuca vivipara																	80	11	2
Kobresia myosuroides																		3	2
No. of species, range	5-8			13-18	11-23		15-26			17-30		19-26		16-30			11-27		
No. of species, average	6			16	17		19			23		22		24			18		
Summa F%, range	230-380			675-910	505-1255		690-1170			520-1585		710-1025		465-1125			440-870		
Summa F%, average	297			790	866		940			1162		881		868			673		

Table 5. Ass. Phippsietum algidae-concinnae, phanerogams. Subass. typicum (Z2), subass. oxyrietosum digynae (B4), and two communities (B5, Z3).

level ground, a clear zonation of one to a couple of metres wide vegetation belts are seen. Where the snow remains until late July or beginning of August no phanerogams are seen on the moist to wet ground, which is more or less covered by organic crust. Then comes a zone with very scattered, sterile *Phippsia algida*, and a zone

	B4				B5			
Variant/community	4.1	4.2			5.1	5.2		
No. of relevés	*3*		*6*	*9*	*2*		*5*	*7*
F% (F) or average F% (av.)	av.	F	av.		av.	F	av.	
Analysis no.		144				199		
Distichium capillaceum	13	80	40	*7*	45	50	54	*6*
Cladonia pyxidata	33	70	23	*5*	20	10	38	*6*
Pohlia cruda	27	10	2	*2*	45	10	24	*6*
Cetrariella delisei	7	30	5	*2*	45	40	48	*6*
Polytrichastrum alpinum	97		12	*3*	25	30	6	*2*
Bartramia ityphylla	23	10	3	*4*			8	*2*
Ceratodon purpureus	23			*1*			4	*2*
Polytrichum juniperinum		70	17	*2*	55		38	*5*
Stereocaulon paschale			17	*2*	20		38	*4*
Ditrichum flexicaule			22	*3*		50	32	*3*
Sanionia uncinatus		10	2	*1*	10		2	*2*
Amphidium lapponicum			6	*1*		20	6	*2*
Pohlia obtusifolia	30		17	*2*				
Anthelia juratzkana	33		7	*2*				
Conostomum tetragonum	23		5	*2*				
Gymnomitrion concinnata	13		8	*2*				
Psilopilum cavifolium	3		2	*2*				
Tortella fragilis					30	60	62	*7*
Hypnum revolutum					15	10	14	*3*
Isopterygiopsis pulchellum					5		4	*2*
Fissidens osmundoides							6	*2*
Saelania glaucescens							4	*2*
No. of species, range	5-10	3-10			12	8-15		
No. of species, average	8	7			12	11		
Summa F%, range	240-540	140-390			320-550	330-700		
Summa F%, average	390	275			435	464		

Table 6. Ass. Phippsietum algidae-concinnae subass. oxyrietosum digynae (B4) and *Salix arctica-Saxifraga cernua* comm. (B5), cryptogams.

with more, now fertile *Phippsia*, in the next zone accompanied by some sterile *Luzula confusa*, and single, mostly sterile *Saxifraga hyperborea, S. foliolosa*, or *S. tenuis*. Here, the soil is covered by a carpet of organic crust, mosses, *Stereocaulon*, and sometimes *Cetrariella delisei*. Seven relevés from the latter zone are representatives of the association (Fredskild 1996, anal. 58, 106, 108, 111, 112, 121, 133), which is common on the poor soil west of the river, markedly less frequent on the more rich soil east of it, represented by only one relevé (anal. 58). The few *Salix arctica* occurring in some relevés are all tiny, the size of *Salix herbacea*, and sterile.

Characteristic of the subassociation, which is widespread in Northeast Greenland, is the all dominating *Phippsia algida*, the low diversity of phanerogams, and the absence of many species frequent in the other associations of the suballiance, e.g. *Cerastium articum, Draba adamsii, D. lactea, Oxyria digyna*, and *Trisetum spicatum*. The type relevé (Table 5, anal. 106) is from a less than 5° south facing slope next to a snow drift, c. 90 m a.s.l. All phanerogams were sterile on July 27. The soil is covered by organic crust, mosses and lichens, i.a. *Stereocaulon alpinum* and *Cladonia pocillum*.

4.2.2. Late-moderate snowbed vegetation

Subass. oxyrietosum digynae subass. nov. (B4)
Differential species *Oxyria digyna, Draba adamsii*, and *Cerastium arcticum*.

Typical variant var. nov. (B4.2)
Seven middle altitude relevés from Traill Ø (Tables 5-6) represent this subassociation of Phippsietum algidae-concinnae. They are mainly found on horizontal or only slightly sloping ground, often with large polygons. The moss cover is light to medium. Judging from the few soil analyses with these species pH is weakly acid (Fig. 11), confirmed by the field measurements (pH 5.1-6.4, 6 anal.), and the conductivity is low. The moss cover is light to medium, *Distichium capillaceum* being the most frequent (Table 6).

The type relevé (Tables 5-6, anal. 144) is from "an almost plain terrain with very flat, large polygons with vegetation all over. Intermediate as to humidity. *Salix arctica* less conspicuous, mostly very small specimens" (Sørensen s.a.). 550 m a.s.l., pH 6.1. In one relevé only: *Draba crassifolia, Scorpidium turgescens, Brachythecium groenlandicum, Bryoerythrophyllum recurvirostre*, and in the type relevé *Fissidens viridulus* (10%).

As far north as 77°30' N late snowbeds with i.a. *Phippsia algida, Draba adamsii, Saxifraga tenuis*, and *S. hyperborea* may represent the subassociation (Bay & Fredskild 1991). *Phippsia* often characterizes the very latest snow free part. In North

Greenland *Saxifraga hyperborea, S. tenuis,* and *Phippsia algida* are characteristic for the late snowbeds (Bay 1992).

Variant of *Cerastium regelii* var. nov. (B4.3)
Of four relevés from the outer coast of Hold with Hope, the two are from the small isle of Hollænder Ø with 89 m the highest point, the other two from the small peninsula Kap Broer Ruys, likewise from the lowland. A typical relevé (Table 5, anal. 16) is from the latter loc. on slightly N-facing, moist, gravelly ground with stones and boulders. In a footnote to the anal. Sørensen mentions on *Festuca ovina (= F. brachyphylla)* "as well the normal as the viviparous form appear, the latter the less common", indicating the occurrence of *F. vivipara*.

The two species: *Cochlearia groenlandica* and *Cerastium regelii* point to a longer snowcover and a more moist ground than in the typical variant.

Variant of *Saxifraga oppositifolia* var. nov. (B4.1)
The only moderate snowbeds from the interior are 3 quite different relevés, the two of which are from an E-facing slope 600 m a.s.l. Below a large snow drift first came a very late snowfree *Phippsia algida* snowbed with a few *Luzula confusa* (not incl. in Table 5), then a late snowbed (in Table 5), dominated by *Phippsia algida, Luzula confusa, Saxifraga oppositifolia,* and *Salix arctica*, whereas an earlier snowfree vegetation (in Table 5), with *Potentilla hyparctica, Draba lactea,* and *Oxyria digyna*, but without *Phippsia* was found next to the snow drift and the snowbeds. The third relevé was called a *Ranunculus pygmaeus* snowbed (275 m). Common is the dominating *Polytrichastrum alpinum*. In one relevé only: *Pogonatum dentatum, Polytrichum piliferum.*

Salix arctica-Saxifraga cernua community (B5)
Subtype of *Minuartia stricta* (B5.2)
Seven low to middle altitude relevés from Traill Ø are from plains, only exceptionally slightly south sloping, all termed "*Salix* plain". Typically, they form a belt between late snowbeds and different heath types, reflected in the all dominating *Salix arctica* and in many of the differential species against Phippsietum subass. typicum: *Carex misandra, Pedicularis hirsuta, Minuartia biflora, Carex nardina, Papaver radicatum, Dryas octopetala, Melandrium apetalum, Cetrariella delisei, Tortella fragilis, Polytrichum juniperinum* (Tables 5-6). pH is 5.3-6.7 (7 anal.).

A typical relevé (Tables 5-6, anal. 199) is from an almost level basalt plateau with many grey, epigaeic lichens, fairly late free of snow, 200 m a.s.l., pH 6.5. In one relevé only: *Braya purpurascens, Carex bigelowii, C. scirpoidea, Pedicularis flammea, Saxifraga aizoides, Blepharostoma trichophyllum, Hypnum bambergeri, Polytrichum strictum, Warnstorfia exannulata*, and, in the typical relevé, *Pohlia proligera* (30%) and *Calliergon trifarium* (40%).

Subtype of *Taraxacum arcticum* (B5.1)
Five middle altitude relevés from Traill Ø differ in the frequent occurrence of *Taraxacum arcticum* and *Trisetum spicatum* (in 4 of 5 rel.), and, less frequent, of *Ranunculus pygmaeus, Draba subcapitata,* and *Ranunculus sulphureus*. They are found on almost level ground, pH 5.5-6.7. A typical relevé (anal. 161) is from a "snowbed" 600 m a.s.l., pH 6.4. In one relevé only: *Poa pratensis* ssp. *alpigena, Tortula ruralis, Mnium thomsonii, Encalypta rhabdocarpa, Myurella julacea.*

Subtype of *Potentilla hyparctica* (B5.3)
Four lowland, outer coast relevés from Hold with Hope are more poor in the thermophilous *Trisetum spicatum* and *Minuartia biflora*, whereas the northern *Potentilla hyparctica* is more frequent. A typical relevé (Table 5, anal. 43) is from a "moist *Polygonum viviparum* fellfield, gravelly-stony, covered by snow during winter, most species weakly developed and, with the exception of *Polygonum vivi-*

parum, only sparsely fruiting". In one relevé only: *Arenaria pseudofrigida, Carex rupestris, C. scirpoidea, Melandrium affine*, and in the typical relevé *Campanula uniflora* (20%), *Potentilla hookeriana/nivea* (10%), and *Ranunculus glacialis* (20%).

Alopecurus alpinus-Saxifraga tenuis community (Z3)

With one exception all analyzed snowbed vegetations at Zackenberg are found below 100 m, usually forming narrow belts on south slopes between heath vegetations above, and late melting snowdrifts covering almost barren ground below. 11 relevés represent the community (Fredskild & Bay 1993, anal. 49-51, 53, 54, 56, 60-63, Fredskild 1996, anal. 134), three of these illustrating the zonation on a southwest slope of a moraine 33 m a.s.l. The abrassion flat on its top has a very open *Dryas-Carex nardina* heath, followed by a belt of *Cassiope* heath with *Dryas* on the uppermost part of the slope. Here, the snow melts in the beginning of July. Then follows an early snowbed with many *Poa arctica* and *Trisetum spicatum*, and with some *Oxyria digyna* (anal. 49), then a somewhat later snowfree belt with many *Alopecurus alpinus, Luzula confusa*, and some *Sagina intermedia* (anal. 50), and finally an even later snowfree belt with the same graminoids, yet fewer specimens, and more *Sagina intermedia* and *Ranunculus pygmaeus* (anal. 51). Further downslope the mainly wind-deposited sand is almost barren. Common to the three belts is the dominating *Salix arctica* and the frequent *Alopecurus alpinus, Poa arctica,* and *Luzula confusa. Alopecurus alpinus* is characteristic of the community. *Poa arctica, Potentilla hyparctica,* and *Stellaria longipes* are markedly more frequent. In one relevé only: *Equisetum variegatum* and *Saxifraga nivalis. Saxifraga hyperborea* (but no *S. rivularis*) was found in two relevés. A typical relevé (Table 5, anal. 56) is from a 4° NE-facing slope, 36 m a.s.l.

Generally, the snowbed vegetations here have a lower diversity and degree of cover than in the southern area, and graminoids are more prominent. The moss cover is large, whereas the cover of organic crust varies between 0 and 100%. By far most species found in the 11 relevés are circumgreenlandic or high arctic, only two, *Ranunculus pygmaeus* and *Minuartia biflora* are southern, having their N-limit at c. 79°. They both characterise the early snowbeds of the community as far north as Nørre Mellemland (78°30' N) and Lambert Land (79°10'N), with e.g. *Trisetum spicatum, Festuca hyperborea, Taraxacum arcticum, Salix arctica, Luzula confusa,* and *L. arctica* (Bay & Fredskild 1991).

	D9			Z4			D10	
Area	HH			Z			HH	
Elevation, m	< 100			35-80			< 100	
No. of relevés			*5*			*5*		*4*
F% (F) or average F% (av.)		av.			av.		av.	
Analysis no.	36			105				
Luzula confusa	40	52	*5*	50	64	*5*	43	*4*
Polygonum viviparum	100	86	*5*	70	64	*4*	90	*4*
Salix arctica	10	24	*4*	50	76	*5*	73	*4*
Salix herbacea	100	84	*5*	100	100	*5*	30	*2*
Poa arctica	100	38	*3*	80	58	*4*	100	*4*
Potentilla hyparctica	40	32	*5*		20	*2*	15	*4*
Carex bigelowii	100	60	*3*		50	*2*	60	*3*
Stellaria longipes s.l.	70	16	*2*		2	*1*	78	*4*
Ranunculus nivalis	100	26	*2*	30	16	*4*	8	*1*
Luzula arctica		4	*2*	10	10	*4*	3	*1*
Pedicularis hirsuta	10	12	*3*	+	+	*1*	10	*1*
Carex lachenalii		30	*3*					
Draba lactea		10	*3*					
Juncus biglumis		20	*2*					
Koenigia islandica		8	*2*					
Ranunculus pygmaeus	20	34	*5*	70	26	*4*		
Minuartia biflora	50	56	*5*	50	40	*3*		
Trisetum spicatum		22	*2*	30	42	*5*		
Silene acaulis		32	*4*	20	6	*2*		
Erigeron humilis	10	40	*3*		10	*1*		
Taraxacum arcticum		10	*1*	+	2	*3*		
Festuca rubra		22	*2*		6	*1*		
*Festuca brachyphylla/*s.l.		6	*2*		4	*1*		
Cerastium arcticum	40	28	*2*		2	*1*		
Equisetum variegatum		10	*2*	90	18	*1*		
Saxifraga cernua		10	*2*		+	*1*		
Vaccinium microphyllum	10	2	*1*				75	*3*
Ranunculus sulphureus	10	24	*3*				3	*1*
Dryas octopetala		2	*1*				10	*1*
Campanula uniflora		4	*2*				3	*1*
Oxyria digyna				100	54	*5*		
Alopecurus alpinus				20	28	*4*		
Cassiope tetragona				20	16	*3*		
No. of species, range	12-25			12-20			9-13	
No. of species, average	18			15			11	
Sum of F%, range	560-750			510-800			510-1170	
Sum of F%, average	625			694			846	

Table 7. Ass. Luzulo-Salicetetum herbaceae (D9, Z4) and a related group (D10), phanerogams.

4.2.3. Moderate snowbed vegetation

One of the 18 ecosystems described by Sørensen in Sørensen & Seidenfaden (1937) is called: "Wet, sheltered sunny Slopes and Patches. - Meagre hygrophilous Herb Mats", also termed "meagre herb mats (luxuriant snow-patch vegetations)". Species, almost exclusively associated with this ecosystem, are *Salix herbacea, Minuartia biflora, Ranunculus pygmaeus, R. sulphureus, R. nivalis, Erigeron humilis, Trisetum spicatum, Taraxacum arcticum,* and *Epilobium arcticum.* With the exception of *Epilobium arcticum* they are all characteristic of or occur in Group D, which according to Sørensen (s.a.) belongs to Nordhagens Cassiopeto-Salicion herbaceae. "The All. is only represented in the outer coast zone. It is primarily characterized by *Salix herbacea,* and besides by *Ranunculus sulphureus.* It falls into a species rich, more luxuriant type (f) with *Minuartia biflora, Potentilla emarginata (= hyparctica), Ranunculus pygmaeus,* and a species poor, more meagre type (g) with *Pedicularis hirsuta* and *Carex rigida (= bigelowii).* In Scandinavia this vegetation type would be considered a snowbed vegetation. In Northeast Greenland it is developed on fairly favourable, often south exposed sites, most likely with a moderate snow cover". Probably the soil of "type f" is more basic than "type g", but no pH measurements are available. The mentioning of *Pedicularis hirsuta* as characteristic of type g seems odd, as it occurs in only one (F=20%) of the four relevés, against three of five in type f.

Ass. Luzulo-Salicetum herbaceae Daniëls & Fredskild ass. nov. (D9, Z4)
Regional character species is *Salix herbacea*; differential species are *Luzula confusa* and *L. arctica.* The variation is described on the level of variants.

Typical variant var. nov. (D9)
Five lowland relevés from Hold with Hope are from level or only slightly sloping ground at the transition between slopes and plains. No moss analyses are at hand, but especially *Polytrichum* sp. seems to be frequent. A very mossy relevé is heavily grazed by geese. Other cryptogams mentioned are *Anthelia juratzkana, Stereocaulon,* and *Cetrariella delisei.* The type relevé (Table 7, anal. 36) is from a slope below a "Flag" (see below: Koenigio islandicae-Saginetum intermediae, C7), rich in mosses.

Daniëls (1982 pp. 63-64) discusses the two low arctic Greenland alliances of the order Salicetalia herbaceae Br.-Bl.: Cassiopo-Salicion herbaceae and Ranunculo-Oxyrion digynae. The former consists of "low snow bed communities on poor soil poor in higher plants, but rich in cryptogams". He states that "As noticed by many investigators, the differentiation of the alliance as against the phanerogam rich Ranunculo-Oxyrion (and the Stellaria-Oxyrion) is rather difficult and arbitrary. The gradual increase of the phanerogams (and decrease of cryptogams.....) coincides with the gradually decreasing thickness and duration of the snow cover". Consequently, Daniëls (1994), following Dierssen (1992) united these two alliances into one: Salicion herbaceae.

With the exception of *Salix herbacea* all common species in this association either

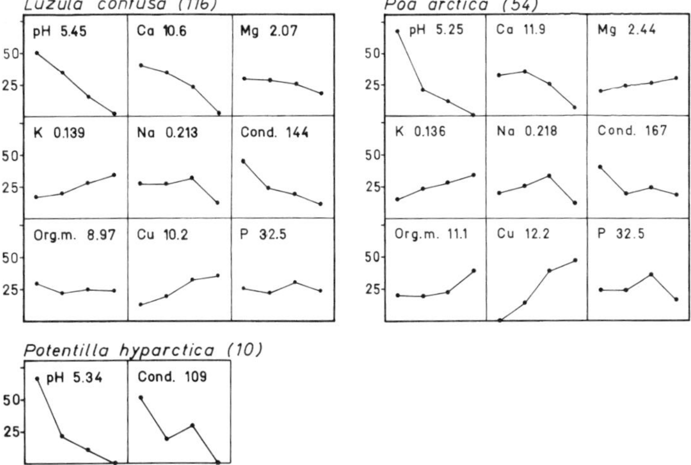

Fig. 12. Soil characteristica for three species common in ass. Luzulo-Salicetum herbaceae.

have a very wide phytosociological amplitude (*Luzula confusa, Polygonum viviparum, Salix arctica, Silene acaulis*) or are mainly or exclusively occurring in Saxifrago-Ranunculion nivalis (*Potentilla hyparctica, Minuartia biflora, Ranunculus pygmaeus*). Consequently, this middle arctic association, which can be considered the northernmost, coastal, impoverished outlier of the southern "herb slope" vegetations, may rather be placed under Drabo-Cardaminenion bellidifoliae of Ranunculo-Oxyrion. Character species of the association is Salix herbacea.

Ecologically, the most frequent species are either ubiquitous or mainly occurring on acid soil with low conductivity (Figs 6, 7, 12). In one relevé only: *Cerastium regelii, Draba nivalis, D. subcapitata, Kobresia myosuroides, Equisetum arvense*, and *Melandrium apetalum.*

The four group g relevés from Hold with Hope (Table 7, D10) differ much from group f (D9), especially in their low phanerogam diversity. The dominating species are phytosociologically widespread or, at least, occur in several other plant communities (*Stellaria longipes* s.l.). Only *Potentilla hyparctica* in four, and *Salix herbacea* in two relevés show affinity to "Alliance D". As the relation of group g with groups d, l and k is almost the same (QS = 41-43) as with group f (QS = 45), and as further the four relevés are mutually quite different, the group is disregarded.

Variant of *Oxyria digyna* var. nov. (Z4)

East of the Zackenberg river *Salix herbacea* is very rare, and only one snowbed with this species was analyzed (Fredskild 1996, anal. 55), whereas such snowbeds are not infrequent west of the river. Four lowland relevés are from slightly tilting south facing slopes, upwards replaced by *Cassiope* heaths or herb-slopes, downwards by later snowbed vegetations. *Salix herbacea* is characteristic; differential species against Luzulo-Salicetum herbaceae typical variant are *Oxyria digyna* and *Alopecurus alpinus; Trisetum spicatum* is more frequent (Table 7). *Salix herbacea* has its known N-limit at 76°46'N, but as far north as Ostenfeld Land (75°18'N) and Hochstetter Forland (75°43'N) *Salix herbacea* snowbeds rich in species have been found (Bay & Fredskild 1990). The ground is almost covered by mosses. Lichens, sometimes *Peltigera*, but fewer *Stereocaulon*, are mostly frequent.

A typical relevé (Table 7, anal. 105) is from a 3 m wide belt on a less than 5° south facing slope of a moraine, 80 m a.s.l., upwards replaced by a *Cassiope* heath, downslope by an *Anthelia* snowbed with *Luzula confusa* and *Carex subspathacea*, bordering a lake and grazed by geese. In one relevé *Cardamine bellidifolia, Phippsia algida, Saxifraga oppositifolia*, and in the typical relevé *Eriophorum triste* (10%).

The statement by Sørensen (s.a.) that the "Alliance D" is only represented in the outer coast zone seems to be at variance with Seidenfaden & Sørensen (1937) according to which the "Wet, sheltered sunny slopes and patches. - Meagre hygrophilous herb mats" are developed both in the inner fjord and the outer coast zones on all kinds of soil, and especially *Ranunculus pygmaeus* is here mentioned as being characteristic for the inner fjord zone.

North of 75°N these early, species rich snowbeds are only found in the inland, mainly on protected S-facing slopes. They may be representatives of a northern association or subassociation, named by *Ranunculus pygmaeus* and *Alopecurus alpinus*, and closely related to Luzulo-Salicetum herbaceae, but no analyses are at hand. Charcteristic species are *Taraxacum arcticum, Minuartia biflora, Trisetum spicatum, Ranunculus pygmaeus, R. nivalis, Alopecurus alpinus, Potentilla hyparctica*, and *Draba arctogena*. As far north as Lambert Land (79°10'N) an early snowbed dominated by *Ranunculus pygmaeus*, here at its northernmost Greenland locality, and *Alopecurus alpinus* was found on an east slope 150 m a.s.l. (Bay & Fredskild 1991). Other species were *Cerastium arcticum, Draba alpina, D. arctogena, D. lactea, D. subcapitata, Minu-*

artia biflora, Phippsia algida, Potentilla hyparctica, Ranunculus nivalis, R. sulphureus, Saxifraga cernua, S. hyparctica, and *S. nivalis.*

4.3. Middle-arctic, species rich communities on wet ground ("Flag vegetations")

Sørensen (1942) discusses the icelandic "flag vegetations", originally described by Hansen (1930). These vegetations, characterized by tiny forbs (*Koenigia islandica, Sedum villosum, Cerastium caespitosum, Sagina intermedia, S. nodosa*) and graminoids (*Juncus biglumis, J. triglumis,* and *Agrostis stolonifera*) occur on moist to wet, fine grained, mainly clayey minerogenous soil, often with seeping water. Sometimes it is slightly hummocky. Three associations were suggested, of which one, the low arctic Koenigio-Sedetum villosi shows affinities to some high arctic Northeast Greenland communities which Sørensen, in Seidenfaden & Sørensen (1937) had termed "Ecosystem: Denuded Bogs". However, in Sørensen (1942) a new Alliance: Koenigio-Microjunceon (arcticum) which should include Koenigio-Sedetum villosum in North Iceland as well as the Northeast Greenland vegetations grouped under C, was suggested. According to the code (Barkman & al. 1986) the alliance is invalid. The East Greenland representatives are considered an association of Saxifrago-Ranunculion nivalis.

Group C is by Sørensen (s.a.) described as: "Characteristic species are *Koenigia islandica, Melandrium apetalum, Colpodium vahlianum* and, moreover, *Juncus biglumis*, which is not confined to the group. Typically, this vegetation type occurs only at the outer coast, whereas it is fragmentaric, found only at high altitudes in the middle and interior Fjord zone. Depending on soil conditions it can be divided into an acidophilous type (d) with *Poa arctica, Cardamine bellidifolia, Ranunculus*

Variant	C7			Z5			C8	
Area	HH			Z			T	
Elevation, m	<100			2-600			575	
No. of relevés			*7*			*7*		
F% (F) or average F% (av.)	F	av.		F	av.		F	F
Analysis no.	26			98			179	180
Polygonum viviparum	90	81	*7*	90	67	*7*	100	100
Salix arctica	20	50	*7*	90	81	*7*	90	10
Juncus biglumis	80	87	*7*	70	39	*6*	90	100
Koenigia islandica	100	76	*7*	30	16	*4*	100	90
Saxifraga cernua	80	53	*7*	60	33	*7*		95
Carex misandra		54	*6*	80	14	*4*	75	75
Festuca hyperborea	70	64	*6*	30	24	*5*	15	60
Saxifraga tenuis	70	57	*7*	+	19	*5*		10
Sagina caespitosa	90	34	*4*					
Ranunculus glacialis	70	19	*4*					
Draba bellii		11	*3*					
Draba subcapitata	10	3	*2*					
Eutrema edwardsii		11	*2*					
Papaver radicatum		9	*2*					
Minuartia rubella	10	23	*4*	30	10	*3*		
Saxifraga platysepala		17	*2*	10	16	*4*		
Saxifraga hirculus		14	*1*	20	21	*5*		
Pedicularis hirsuta		10	*1*	+	10	*5*		
Draba adamsii		7	*2*		3	*2*		
Minuartia biflora		7	*2*		1	*1*		
Taraxacum arcticum		1	*1*		1	*1*		
Alopecurus alpinus	80	61	*6*	90	41	*4*		
Saxifraga foliolosa	100	61	*6*	20	14	*2*		
Deschampsia brevifolia	20	7	*3*	50	10	*2*		
Juncus castaneus	20	6	*2*		+	*1*		
Arctagrostis latifolia		10	*1*		1	*1*		
Draba alpina	30	9	*2*	50	7	*1*		
Cerastium regelii		4	*1*	20	3	*1*		
Luzula arctica	10	61	*7*	50	57	*7*	85	80
Stellaria longipes s.l.	100	67	*7*	30	46	*7*	15	60
Dryas octopetala/sp.		3	*2*	40	29	*6*	25	
Carex maritima	90	40	*4*		9	*2*		100
Carex bigelowii		7	*1*		1	*1*		10
Eriophorum triste		6	*1*	30	6	*3*		
Equisetum arvense	90	74	*6*	70	23	*2*		
Draba lactea	70	57	*7*	20	20	*6*	65	5
Cerastium arcticum	20	40	*7*	70	29	*5*	5	15
Sagina intermedia	30	43	*6*		27	*6*	25	
Melandrium apetalum	10	21	*5*	10	7	*4*	30	20
Carex nardina		1	*1*	20	3	*1*	30	10
Luzula confusa	40	24	*4*		29	*6*		25
Saxifraga oppositifolia		36	*5*		40	*4*	70	70
Cardamine bellidifolia	20	11	*3*		6	*2*		5
Poa arctica	10	14	*3*	20	33	*6*		
Silene acaulis		24	*2*	10	7	*5*		
Potentilla hyparctica		10	*3*		6	*1*		
Ranunculus sulphureus	10	9	*3*					
Colpodium vahlianum	90	30	*4*				100	50
Poa glauca				10	11	*4*		
Kobresia myosuroides				20	20	*3*		
Saxifraga nivalis					6	*2*		
Saxifraga caespitosa					1	*2*		
Festuca baffinensis				+	1	*2*		
Pedicularis flammea				+	1	*2*		
Equisetum variegatum					3	*2*		
Juncus triglumis				10	1	*1*	95	25
Carex atrofusca					1	*1*		10
Carex rupestris				70	17	*2*	35	
Carex capillaris					4	*1*		5
Carex saxatilis								
Kobresia simpliciuscula							10	5
Saxifraga aizoides							100	70
Epilobium arcticum							100	10
Minuartia stricta							30	35
No. of species, range	23-33			19-34			33	27
No of species, average	27			26				
Summa F%, range	1050-1600			600-1260				
Summa F%, average	1437			899				

Table 8. Ass. Koenigio-Saginetum intermediae, phanerogams.

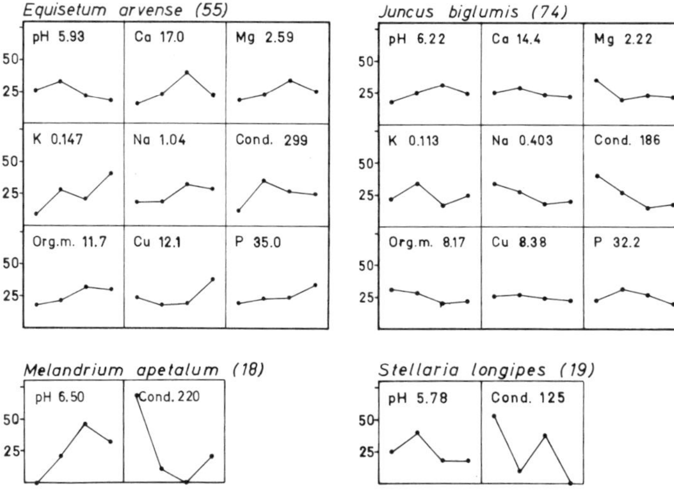

Fig. 13. Soil characteristica for four species common in ass. Koenigio-Saginetum intermediae.

glacialis, and a basophilous type (e) with *Epilobium arcticum, Juncus triglumis, Saxifraga aizoides, Minuartia stricta*".

Ass. Koenigio-Saginetum intermediae Daniëls & Fredskild ass. nov. (C7-8, Z5)
The association is characterized by *Sagina intermedia, Koenigia islandica, Festuca hyperborea*, and species such as *Deschampsia brevifolia, Saxifraga hirculus*, and *S. platysepala*.

Variant of *Ranunculus glacialis* var. nov. (C7)
Seven relevés are from the outer coast lowland at Hold with Hope (Table 8). According to the thorough description in Seidenfaden & Sørensen (1937) and the lapidaric field notes they are from belts on level or only slightly sloping ground below snow drifts, usually wet throughout the summer as a result of seeping melt water, but sometimes drying out. Occasionally, hummocks with e.g. *Deschampsia brevifolia* and *Carex misandra* occur. The average number of species (27) is the highest in any association in the 73°N area. This also holds for the average sum of F%, yet not for the cover, as most species are very small. In this area *Alopecurus alpinus*, with the exception of two relevés, only occurs in this association. Likewise, *Draba bellii*, occurring in three relevés here, is only found in one relevé of Saxifrago-Kobresietum simpliciusculae var. typicum (E11) and two of the three relevés of the *Taraxacum phymatocarpum-Poa abbreviata* community (N41). Apart from in one circle of twenty in only one relevé of the latter community *Colpodium vahlianum* only appears in this association, and *Deschampsia brevifolia* is occurring exclusively in the association. Beyond doubt all, or at least by far the major part, of the plants termed *Festuca brachyphylla* in the "Flag vegetation" analyses by Sørensen are actually *F. hyperborea*, which was not realized as a separate species until later (Holmen 1952). As everywhere in Northeast and North Greenland this species characterises just these communities, as confirmed by the analyses from Zackenberg and observations further north, *Festuca hyperborea* is used in Table 8.

Judging from the ecological demands of the species (Fig. 13) the soil is fairly rich, with pH 6-7. No information other than "Flag" is given on the type relevé (Table 8, anal. 26). In one relevé only: *Campanula uniflora, Cochlearia groenlandica, Oxyria digyna, Phippsia algida.*

Variant of *Saxifraga hirculus* var. nov. (Z5)
At Zackenberg some of the vegetations with the highest diversity are found 400-600 m a.s.l. on the southern slope of Aucellabjerg, kept wet throughout the summer by oozing meltwater from a permanent snowdrift 600-700 m a.s.l. The ground is stony clay, locally moving and

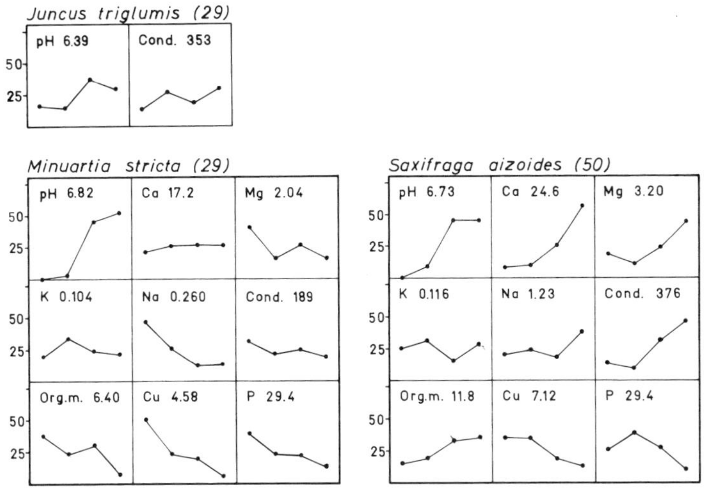

Juncus triglumis (29)
pH 6.39
Cond. 353
Minuartia stricta (29)
pH 6.82
Ca 17.2
Mg 2.04
K 0.104
Na 0.260
Cond. 189
Org.m. 6.40
Cu 4.58
P 29.4
Saxifraga aizoides (50)
pH 6.73
Ca 24.6
Mg 3.20
K 0.116
Na 1.23
Cond. 376
Org.m. 11.8
Cu 7.12
P 29.4
50
25

Fig. 14. Soil charcteristica for three species common in ass. Koenigio-Saginetum intermediae var. of *Saxifraga aizoides.*

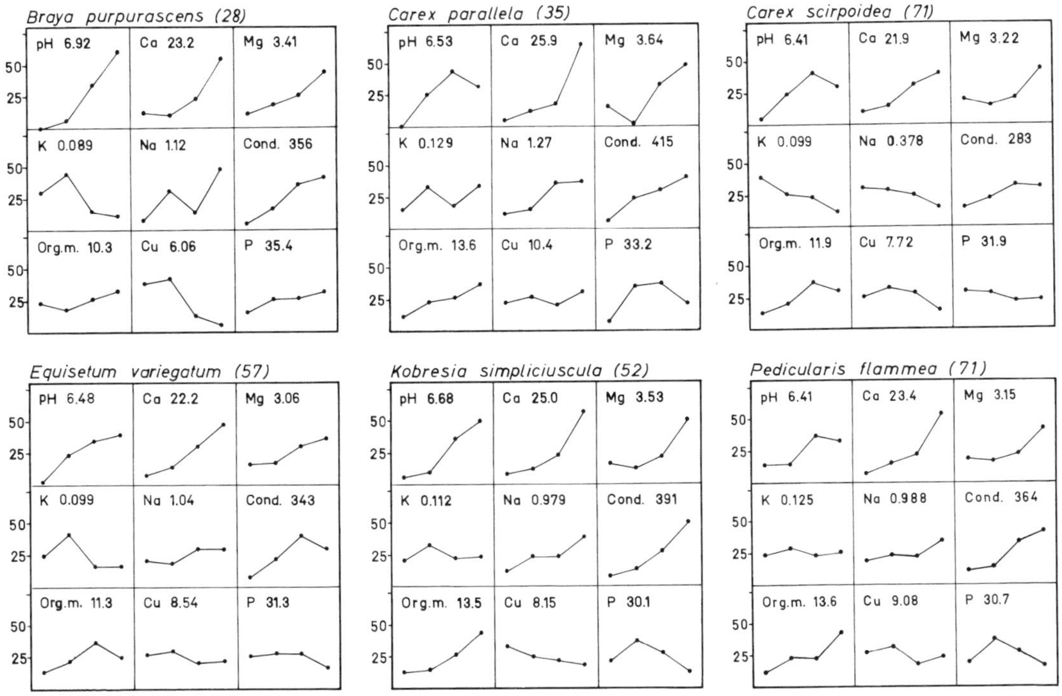

Braya purpurascens (28)
pH 6.92
Ca 23.2
Mg 3.41
K 0.089
Na 1.12
Cond. 356
Org.m. 10.3
Cu 6.06
P 35.4
Carex parallela (35)
pH 6.53
Ca 25.9
Mg 3.64
K 0.129
Na 1.27
Cond. 415
Org.m. 13.6
Cu 10.4
P 33.2
Carex scirpoidea (71)
pH 6.41
Ca 21.9
Mg 3.22
K 0.099
Na 0.378
Cond. 283
Org.m. 11.9
Cu 7.72
P 31.9
Equisetum variegatum (57)
pH 6.48
Ca 22.2
Mg 3.06
K 0.099
Na 1.04
Cond. 343
Org.m. 11.3
Cu 8.54
P 31.3
Kobresia simpliciuscula (52)
pH 6.68
Ca 25.0
Mg 3.53
K 0.112
Na 0.979
Cond. 391
Org.m. 13.5
Cu 8.15
P 30.1
Pedicularis flammea (71)
pH 6.41
Ca 23.4
Mg 3.15
K 0.125
Na 0.988
Cond. 364
Org.m. 13.6
Cu 9.08
P 30.7
50
25

Fig. 15. Soil characteristica for six species common in ass. Saxifrago-Kobresietum simpliciusculae.

Carex atrofusca (15)
pH 6.34
Cond. 712
Eriophorum callitrix (13)
pH 6.08
Cond. 755
Euphrasia arctica (23)
pH 6.26
Cond. 372
Eutrema edwardsii (12)
pH 6.85
Cond. 301
Pedicularis lapponica (40)
pH 6.18
Cond. 313
Saxifraga nathorstii (27)
pH 6.79
Cond. 441
50
25

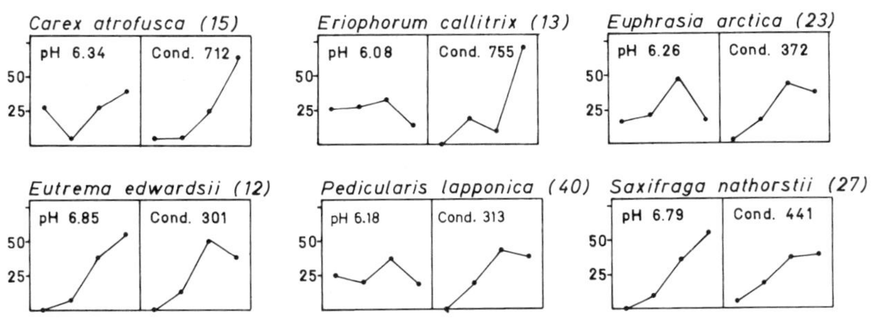

Fig. 16. pH and conductivity for six species common in ass. Saxifrago-Kobresietum simpliciusculae.

bare, but mostly stabilized and with a dense cover of moss, organic crust, and "plates" of *Nostoc*. Four relevés represent the variant (Fredskild & Bay 1993, anal. 97-99, 102). One of these (Table 8, anal. 98) has the highest number of phanerogams (34) of all in the present investigation. This relevé is from a 13° S-slope with seeping water 420 m a.s.l., by the end of July moist to very wet. As a result of the analysis method with circles at regular intervals along a fixed line, the rather dense cover of phanerogams includes some small, drier spots with *Dryas, Carex nardina*, and *Kobresia myosuroides*, not belonging to the ass. If these are disregarded, the variant mainly differs in the more frequent *Saxifraga hirculus* and *Poa arctica*, and the absence of *Sagina caespitosa, Ranunculus glacialis*, and *Draba bellii*. The cover of moss and *Nostoc* is fairly dense.

The other site with representatives of the variant is the c. 1x1½ km wide delta of the Zackenberg river. Above the present high water line the ground consists of pebbles, mainly covered by a one to a few cm thich carpet of organic crust, and many species of mosses and lichens. The different vegetations are arranged in long, one to a few metres wide belts in the former channels and on the intervening longshore bars. In spite of a difference in elevation of only 1-2 dm, the vegetations are quite different. E.g. one of the three relevés is a *Saxifraga hirculus-Salix arctica-Luzula arctica* dominated vegetation (Fredskild 1996, anal. 119), on one side with a meltwater channel from a large snowdrift on a nearby slope, on the other side, only few cm higher up, is a dry, gravelly-stony bar with an open *Dryas-Salix arctica-Silene acaulis* vegetation. A nearby *Sagina intermedia-Salix arctica-Saxifraga platysepala* vegetation (l.c., anal. 122) without *Saxifraga hirculus* is growing on slightly drier ground, as reflected also in the more lichens and very little organic crust. *Alopecurus alpinus*, common in the four high altitude relevés, does not grow in the relevés from the delta.

Community	C6	
Area	H	
Elevation, m	425-600	
Analysis no.	263	250
Eriophorum scheuchzeri	75	50
Poa arctica	20	95
Salix arctica	40	75
Polygonum viviparum	10	100
Juncus biglumis	10	100
Luzula confusa	15	95
Koenigia islandica	15	85
Saxifraga tenuis	5	20
Carex misandra	5	10
Festuca hyperborea	5	10
Ranunculus hyperboreus	100	
Equisetum arvense	90	
Phippsia algida	65	
Carex lachenalii	30	
Saxifraga cernua	10	
Eriophorum triste	10	
Saxifraga oppositifolia	5	
Ranunculus pygmaeus	5	
Potentilla hyparctica		100
Cardamine bellidifolia		85
Hierochloë alpina		40
Carex capillaris		35
Carex rupestris		20
Silene acaulis		5
Polytrichastrum alpinum	40	100
Pohlia obtusifolia	100	
Pohlia cruda	40	
Polytrichum piliferum	40	
Sanionia uncinatus	20	
Nostoc sp.	20	
Oncophorus wahlenbergii		90
Anastrophyllum minutum		90
Campylopus schimperi		80
Aulacomnium turgidum		50
Anthelia juratzkana		40
Distichium capillaceum		40
Amphidium lapponicum		30
Cladonia pyxidata		30
Conostomum tetragonum		30
Meesia uliginosa		20
Scorpidium turgescens		10

Table 9. *Eriophorum scheuchzeri* stands.

In one relevé only: *Arenaria pseudofrigida, Melandrium affine, Poa pratensis* var. *colpodea, Ranunculus affinis, R. nivalis, Trisetum spicatum, Vaccinium microphyllum.*

Variant of *Saxifraga aizoides* var. nov. (C8)
Two relevés from a slightly concave depression with seeping water on a slope at 575 m a.s.l. on Traill Ø (anal. 179 and 180) are from rich soil as indicated by the characteristic species (Fig. 14 and 16). pH is 7.3 and 7.1. Characteristic species is *Epilobium arcticum*, in the 73°N area only registered in these two relevés. Differential species are *Saxifraga aizoides* and *Minuartia stricta*. In anal. 179 only: *Braya*

purpurascens (10%), *Campanula gieseckiana* and *Carex scirpoidea* (both 5%).

In Greenland Koenigio-Saginetum intermediae seems restricted to the east coast between c. 73° and 78°30'N. In the area 74°50'-77°30'N it is fairly frequent, yet covering only very small areas (Fredskild & Bay 1990). These characteristic "black communities", having the highest phanerogam diversity in the area, are found on S-slopes as a zone between snow drifts and grassland communities, or as a belt along permanent meltwater brooklets. During summer the soil is often more or less drying out, the organic crust becoming crispy. Phanerogams cover 5-10%, algae incl. tiny *Nostoc* balls 90-95%. Only exceptionally mosses other than *Anthelia juratzkana* occur. *Juncus biglumis* is dominating, other species being: *Juncus triglumis, Colpodium vahlianum, Carex misandra, Draba lactea, D. adamsii, Cardamine bellidifolia, Ranunculus sulphureus, R. glacialis, Stellaria crassipes, Cerastium arcticum, Sagina intermedia, Saxifraga oppositifolia, S. cernua, S. tenuis, S. foliolosa, Armeria scabra, Potentilla hyparctica, Pedicularis flammea,* and *Koenigia islandica.*

Judging from the descriptions in Schwarzenbach (1961, pp. 132-135) corresponding vegetations can be found at high altitudes (640-1250 m a.s.l.) on nunataks and seminunataks 74°-74½°N. Further northwards the "black communities" are less frequent, yet still very rich in species. An example from Sdr. Mellemland (78°00'N), and another from Nr. Mellemland (78°30'N) is given in Bay & Fredskild (1991). In North Greenland the association is missing, yet a community at Brønlund Fjord in the most continental interior (82°10'N) shows a clear affinity. On the slopes at the east side of lake Klaresø (Fredskild 1973, Fig. 28) a *Juncus biglumis* soc. with many *Stellaria crassipes* and *Colpodium vahlianum,* and scattered *Salix arctica, Cochlearia groenlandica, Alopecurus alpinus, Polygonum viviparum, Carex misandra, Saxifraga oppositifolia, S. cernua, Papaver radicatum, Melandrium triflorum, Draba bellii, D. oblongata,* and *Ranunculus sulphureus* covers a soil with micropolygons (15-20 cm), the moister part of which is covered by a black crust of algae, incl. *Nostoc.* The phanerogams cover less than 25%. Mosses are fairly frequent. pH is 7.9-8.0 (Fredskild unpubl.).

Koenigia islandica, the only therophyte occurring in the high arctic North Greenland, has only been collected four times (Bay 1992) north of 80°, the three times at Brønlund Fjord. The northernmost collection is from a dark coloured, dried out flat at Frigg Fjord (83°10'N, C. Bay, pers. comm.).

Eriophorum scheuchzeri stands (C6)

Two relevés (Table 9) from middle altitude sites at Kap Hedlund may belong to Eriophoretum scheuchzeri Fries 1913 of Scheuchzerio-Caricetea. They are both from moist slopes below snowbeds. The acid ground (pH 5.0 and 4.3, resp.) is clearly reflected in the absence of basophilous species and the low diversity. As to cryptogams, anal. 263 is based on only 5 circles, anal. 250 on 10.

5. All. Caricion atrofusco-saxatilis Nordh. 1943

5.1. Middle-arctic, meso-hygrophytic grassland on rich soil

In Northeast Greenland the separation between the permanently wet fens and the, during the snow melt period wet, but later on drying out grasslands, is easy and often very marked. *Carex stans* (not growing in the 73°N area) and *Eriophorum scheuchzeri* only occur in the fens, whereas *Eriophorum triste, Arctagrostis latifolia, Carex bigelowii,* and *C. misandra* characterize the more hygrophilous, and *Kobresia myosuroides* and *Carex rupestris* the drier grasslands. Especially the grasslands on rich soils are rich in species (Bay 1992, Bay & Fredskild, 1990, 1991). In the northernmost part of the Middle Arctic Tundra Zone *Carex bigelowii* is becoming very rare (N-limit at 78½°, with an isolated occurrence at c. 80½°). Here, *Carex stans*, besides characterizing the fens, may take its role in the grasslands, e.g. on Sdr. Mellemland (77°47'N) and Lambert Land (79°10'N) (Bay & Fredskild 1991). Sørensen (s.a.) did not use the term grassland (but sometimes in the field notes "Græsmyr" = grass mire), writing "Group E represent Caricion atrofuscae-saxatilis (Nordhagen 1943, p. 451), fen vegetation on calcareous, sedimentary rocks, without peat formation, often drying out during summer. The All. is here especially characterized by *Saxifraga aizoides, S. nathorstii*, and besides by *Juncus biglumis, Braya purpurascens, Minuartia stricta,* and further by the following species, also occurring on drier, calcareous soil but usually missing in the other fen communities: *Pedicularis flammea, Kobresia bipartita (= K. simpliciuscula), Carex scirpoidea*. The distribution of this vegetation type is coextensive with the calcareous rocks, i.e. the Middle Fjord area". In Table 10 no less than 11 of the character species, 2 of the regional character species, and 7 of the differential species of Caricion atrofuscae-saxatilis against Schoenion, as mentioned in Nordhagen (1943), occur.

Mosses are mostly frequent. Apart from the ubiquitous species the characteristic species are: *Brachythecium turgidum* and *Catoscopium nigritum*, both almost restricted to this alliance, *Scorpidium turgescens, Campylium stellatum,* and *Hypnum bambergeri* (Table 11).

49 grassland relevés from the 73°N area are grouped into four closely related (QS = 50-56) units, h 11-14, all considered representatives of one ass.: Saxifrago-Kobresietum simpliciusculae.

Ass. Saxifrago-Kobresietum simpliciusculae Daniëls & Fredskild ass. nov. (E11-14, Z6)
Character-species of the association is *Saxifraga nathorstii*. Anal. 344 (Tables 10-11) is nomenclatoric type relevé.

Variant of *Carex saxatilis* (E12) var. nov.
Subvar. of *Carex atrofusca* (E12.2)
Eight relevés from Ella Ø and Ymer Ø, the five of which are below 100 m a.s.l., are from level or only slightly sloping, often minerogenous ground. The moss cover, dominated by *Distichium capillaceum, Ditrichum flexicaule*, and *Hypnum bambergeri* is fairly thin. Characteristic species are *Saxifraga nathorstii*, endemic to East Greenland (70°-75½°N) and the middle-high arctic *Braya purpurascens*. Apart from its occurrence here and in the typical variant (E11) the middle-arctic *Braya linearis* was only found in ass. Arabido holboellii-Caricetum supinae (L37). Preferential species are *Carex atrofusca, C. parallela, Saxifraga aizoides*, and, phytogeographically more wide, *Kobresia simpliciuscula* and *Pedicularis flammea*. The variant resembles Caricetum microglochinis (Nordhagen 1943 p. 455) not only in their many common species, but also in the rich soil and the

Subassociation/community	E11			12.1	E12.2			E13			Z6		14.1	E14.2		
Area	EY			Y	EY			T			Z		Y	EY		
Elevation, m	20-350			40-100	10-400			200-550			25-360		75-300	20-375		
No. of relevés			*12*	*4*		*8*	*12*			*6*			*4*		*15*	*19*
F% (F) or average F% (av.)	F	av.		av.	F	av.		F	av.				av.	F	av.	
Analysis no.	108				344			175			100	101		9		
Polygonum viviparum	100	90	*12*	81	100	84	*12*	100	99	*6*	90	60	95	100	98	*19*
Salix arctica	10	42	*11*	96	95	87	*12*	85	87	*6*	90	50	81	90	63	*19*
Dryas octopetala	100	85	*12*	35	65	29	*10*	100	75	*6*	40	50	48	100	83	*18*
Carex misandra	80	79	*12*	24	95	69	*10*	75	73	*6*	60	30	29	20	42	*14*
Kobresia simpliciuscula	90	82	*12*	14	100	79	*10*	15	29	*3*	40	10	18	40	59	*16*
Eriophorum triste		8	*5*	40	100	80	*11*	35	42	*4*	80	60	21	100	73	*19*
Pedicularis flammea	30	47	*10*	34	35	36	*9*	90	33	*4*	50	60	24		36	*16*
Equisetum variegatum		24	*5*	51	100	88	*11*		3	*3*	30		93	100	76	*18*
Carex rupestris	100	88	*11*	8	35	39	*7*	65	18	*3*	30	10	78	10	27	*12*
Carex parallela		3	*2*	14	95	66	*9*	5	18	*2*	60		33	100	68	*16*
Equisetum arvense		5	*1*	28		39	*7*		50	*3*	30		36		30	*10*
Arctagrostis latifolia		8	*1*		50	51	*5*		15	*1*	100	80	26		7	*4*
Carex capillaris		8	*2*	61			*4*		1	*1*	30	10	1		2	*4*
Juncus biglumis		8	*5*	15	50	21	*7*	60	58	*6*	60	60			12	*6*
Juncus triglumis		5	*3*	8	85	74	*9*	45	37	*4*	70	80			18	*6*
Saxifraga aizoides	65	55	*10*	53	35	35	*11*	85	57	*6*			21	60	39	*13*
Saxifraga oppositifolia	25	65	*11*	39	5	2	*5*	100	90	*6*			54		25	*12*
Saxifraga nathorstii		15	*3*	75	55	40	*10*		14	*3*			40	80	32	*13*
Silene acaulis		8	*6*	10		8	*7*		42	*6*			9		2	*7*
Carex scirpoidea	85	30	*6*	3		16	*5*	100	36	*4*			66		18	*10*
Minuartia stricta		3	*3*	1		1	*2*	10	33	*5*			4		3	*5*
Pedicularis hirsuta		1	*1*	13		1	*3*	10	23	*5*				10	4	*6*
Chamaenerion latifolium	55	18	*5*													
Lesquerella arctica		2	*3*													
Thalictrum alpinum		8	*2*													
Woodsia glabella	35	4	*2*													
Euphrasia frigida	90	20	*4*			26	*3*									
Braya linearis	40	9	*4*			6	*3*									
Melandrium triflorum/affine		4	*3*	14			*1*									
Melandrium apetalum		1	*1*	4			*2*		3	*2*						
Vaccinium microphyllum		2	*3*			9	*2*					60	40	100	63	*17*
Carex atrofusca		2	*1*	14	100	61	*8*				40	60		20	27	*7*
Juncus castaneus		3	*1*	3		9	*2*				60	60			1	*1*
Cassiope tetragona		5	*2*			1	*1*					20	5	10	39	*13*
Betula nana		11	*3*	9		2	*4*						99	80	62	*18*
Kobresia myosuroides	30	62	*10*	20	30	14	*4*						1		5	*5*
Braya purpurascens	10	15	*5*	5	25	42	*8*						10		11	*5*
Tofieldia pusilla	45	18	*4*	3			*2*								39	*11*
Eutrema edwardsii		1	*1*			4	*2*						6	50	4	*8*
Carex nardina	20	16	*5*					15	18	*3*					3	*3*
Luzula arctica		1	*1*						7	*3*	10	10			1	*1*
Pedicularis lapponica		1	*1*										26	40	19	*13*
Carex microglochin				23		13	*3*									
Carex bicolor				11		10	*2*									
Armeria scabra ssp. *sibirica*				5		1	*2*									
Juncus arcticus				34			*2*									
Carex saxatilis				29	25	48	*9*		8	*2*	10	20			11	*4*
Draba lactea				3		1	*2*		19	*4*					1	*1*
Carex rariflora				46			*2*								10	*4*
Carex maritima					60	22	*3*				10					
Eriophorum callitrix					35	9	*3*					20		20	19	*9*
Carex pseudolagopina					40	5	*1*								1	*2*
Carex bigelowii								100	72	*5*	30	80				
Luzula confusa								5	9	*3*						
Saxifraga foliolosa									8	*3*	+	20				
Saxifraga cernua									5	*2*		10				
Poa pratensis ssp. *alpigena*									5	*1*			18			*2*
Koenigia islandica											50	10				
Arctostaphylos alpina													1		1	*2*
Rhododendron lapponicum															5	*3*
No of species, range	15-22			13-25	14-27			19-25			33	27	16-22	16-26		
No of species, average	17			20	21			22					20	21		
Summa F%, range	805-1230			700-1155	615-1760			825-1260					770-1260	765-1495		
Summa F%, average	964			945	1231			1107					1011	1139		

Table 10. Ass. Saxifrago-Kobresietum simpliciusculae, phanerogams.

Subassociation/community	E11			12.1	12.2			E13	14.1	14.2		
No. of relevés			*10*	4		7	*11*		4		15	*19*
F% (F) or average F% (av.)	F	av		av.	F	av.		F	av.	F	av.	
Analysis no.	108				344			137		9		
Distichium capillaceum	70	65	*10*	38	40	44	*10*	80	60	30	54	*18*
Ditrichum flexicaule	20	68	*10*	50	90	61	*10*	80	53	70	61	*17*
Campylium stellatum		18	*5*	35		7	*6*	10	13	100	54	*13*
Bryoerythrophyl. recurvirostre		6	*6*	18	20	10	*6*	70	10		8	*7*
Nostoc sp.	10	23	*6*	10	20	37	*5*	60	18		14	*5*
Tortella fragilis		6	*4*	5		31	*6*	20	18		9	*6*
Meesia uliginosa		10	*5*	25		9	*5*	20	5		11	*5*
Scorpidium turgescens		3	*1*	3		16	*4*	20	23		3	*6*
Encalypta longicollis		16	*4*	3		7	*2*	20	13		9	*4*
Cyrtomnium hymenophylloides	40	14	*5*			7	*3*	70			8	*4*
Physcia muscigena		3	*2*									
Thamnolia vermicularis		1	*1*			1	*1*					
Hypnum bambergeri		49	*8*	35	40	41	*9*		60	90	50	*12*
Calliergon trifarium		7	*1*	8		7	*3*		8	20	9	*5*
Brachythecium binervulum		10	*3*			1	*1*		3		2	*3*
Myurella julacea	30	8	*4*			3	*1*		3		1	*2*
Encalypta procera		6	*3*			1	*1*		3		1	*2*
Schistidium apocarpum		5	*1*	15			*1*		3			*1*
Blepharostoma trichophyllum		1	*1*	3			*1*				12	*6*
Brachythecium turgidum		1	*1*			1	*1*			10	5	*4*
Solorina octospora		2	*1*			1	*1*				1	*2*
Cladonia pyxidata		17	*3*					10		10	7	*4*
Stereocaulon paschale		2	*1*					20			1	*1*
Encalypta rhabdocarpa		7	*2*								3	*2*
Isopterygium pulchellum		2	*2*								3	*3*
Hymenostylium recurvirostre		4	*1*								2	*1*
Bryum wrightii		1	*1*								1	*1*
Distichium inclinatum				28			*2*					
Drepanocladus polycarpus				18		9	*3*		53		1	*4*
Bryum neodamense						4	*1*		28			*2*
Catoscopium nigritum				43	50	26	*6*				11	*5*
Orthothecium cryseum				25			*2*			30	18	*6*
Drepanocladus brevifolius					80	37	*3*				4	*3*
Cinclidium arcticum				3			*1*				11	*5*
Drepanocladus intermedius				13			*1*				14	*2*
Encalypta alpina						3	*1*				1	*1*
Fissidens osmundoides								40	8		1	*5*
Tomenthypnum nitens									25	20	14	*8*
Drepanocladus uncinatus									5		2	*2*
Oncophorus wahlenbergii									3		2	*2*
Pohlia nutans											5	*3*
Brachythecium salebrosum											7	*2*
No. of species, range	6-13			9-13	8-17			14		5-19		
No. of species, average	11			11	11				11	12		
Summa F%, range	190-660			350-550	350-610			530	400-540	300-670		
Summa F%, average	415			460	449				460	510		

Table 11. Ass. Saxifrago-Kobresietum simpliciusculae, cryptogams.

"strange mixture of fen plants with dry ground plants". pH varies between 6.6 and 7.4. The rich soil is confirmed by the very high conductivity of the characteristic species (Figs 14-16). The content of organic matter varies, yet is often fairly high. Among the units under Caricion atrofusco-saxatilis the variant has the highest percentage of graminoids (56% of Summa F%), the lowest of dwarfshrubs (10%). The type relevé (Tables 10-11, anal. 344) is from level ground, 10 m a.s.l., pH 6.6.

In the 73°N area *Carex bicolor* only occurs in one relevé here and in one in the variant of *Carex capillaris* (E12.1), and *Carex microglochin* in two relevés of the first, and one of the last mentioned unit. In one relevé only: *Eurhynchium pulchellum, Funaria hygrometrica, F. polaris, Pottia heimii* var. *arctica, Stegonia latifolia,* and, in the type relevé, *Orthothecium intricatum* (10%).

Subvar. of *Carex capillaris* (E12.1)

Four lowland relevés from Ymer Ø differ in the occurrence of *Carex capillaris* in four, *C. rariflora, Juncus arcticus, Distichium inclinatum,* and *Drepanocladus aduncus* s.l. in two relevés. Information on the sites is given for only one relevé, which is from the bank of a stream. Here, *Carex microglochin* (90%) and *C. bicolor* (45%) are dominating. *Juncus arcticus* has its only two occurrences in the area in this and another of the four relevés.

Typical variant var. nov. (E11)
Eight of the twelve lowland relevés from Ella Ø and Ymer Ø are from 20-100 m a.s.l. The ground is level or slightly sloping, in some cases windexposed and with indications of being (? periodically) snow free during winter, in a few cases confirmed by observations. pH is 6.7-7.5. A typical relevé (Tables 10-11, anal. 108) is from a SW-facing slope, 250 m a.s.l. The variant is found on more drying out soil, as illustrated by the fewer *Equisetum arvense, E. variegatum, Arctagrostis latifolia, Carex atrofusca,* and the missing *Carex saxatilis.* Microtopographical conditions may partly explain the odd mixture of species from fairly moist (e.g. *Kobresia simpliciuscula, Saxifraga aizoides*) and dry ground (*Kobresia myosuroides, Carex nardina, C. rupestris*). In one relevé only: *Draba bellii, Triglochin palustre, Cetraria nivalis, Platydictya jungermannioides.*

Variant of *Carex bigelowii* (E13)
Six relevés from Traill Ø, the one 200, the five 500-550 m a.s.l., differ in the frequent *Juncus biglumis, Pedicularis hirsuta,* and in the many *Carex bigelowii,* in this association occurring only in this variant and in the var. of *Juncus castaneus* (Z6). The five middle altitude relevés are from a "*Dryas* plain", slightly hummocky. pH is 6.2-6.7 (5 anal.). A typical relevé (Table 10, anal. 175) is from an only slightly hummocky, slightly S-facing plain, 525 m a.s.l., pH 6.2. In one relevé only: *Campanula uniflora, Cardamine bellidifolia, Cerastium alpinum, Koenigia islandica, Minuartia biflora, M. rubella, Oxyria digyna, Poa arctica,* and in the typical relevé: *Arenaria pseudofrigida* (10%). The variant has affinity to Koenigio-Saginetum intermediae (C8) as well as to Arctagrostio-Eriophoretum tristis (F20) (QS = 54 and 53 respectively). Moss analysis only available for the 200 m relevé (Table 11, anal. 137).

Variant of *Juncus castaneus* var. nov. (Z6)
Two species rich fen-like communities from Zackenberg are from slopes with seeping water, the one from a c. 20° SW-slope on Aucellabjerg, c. 500 m a.s.l., the other from a less than 5° SE-slope on raised marine deposits, 25 m a.s.l. (Table 10, anal. 100 and 101). *Saxifraga aizoides* and *S. nathorstii*, both rare at Zackenberg, are missing. Not in the table: *Alopecurus alpinus* (20%), *Deschampsia brevifolia* and *Draba alpina* (10%), *Cerastium regelii, Ranunculus sulphureus, Saxifraga tenuis, Stellaria longipes* (+) in anal. 100, *Empetrum hermaphroditum* and *Eriophorum scheuchzeri* (20%), *Festuca hyperborea* and Tofieldia coccinea (10%) in anal. 101.

Variant of *Betula nana* var. nov. (E14)
Subvar. of *Tofieldia pusilla* (E14.2)
15 relevés from Ella Ø-Ymer Ø are from the lowland, only two (250 and 375 m) above 200 m a.s.l., often forming a belt between different heath and fen types. The ground is horizontal or only slightly sloping, sometimes a more or less stabilized solifluction soil, in between hummocky. Dwarfshrubs and occasionally *Eriophorum triste* grow on the hummocks, whereas moist to wet "flag-vegetations" with oozing water are seen between them. The snow cover during winter is often thin. pH of the topsoil is 5.9-7.3 (9 anal.), of the underground 6.8-7.3 (4 anal.). Characteristic species: *Pedicularis lapponica, Eutrema edwardsii, Tofieldia pusilla,* and *Eriophorum callitrix.* Differential species: *Betula nana, Vaccinium uliginosum* ssp. *microphyllum, Cassiope tetragona.*

Among the units of the alliance, this variant has the highest percentage of dwarfshrubs (28% of Summa F%), the lowest of graminoids (36%). *Betula nana* grows in 14, *Vaccinium uliginosum* in 13, *Cassiope tetragona* in 12 relevés, which emphasises the affinity (QS = 55) to the *Betula-Cassiope-Vaccinium* heaths of Saliceto-Cassiopetum tetragonae subass. pyroletosum grandiflorae (G21). On the contrary, the many fen plants show affinity to units of ass. Arctagrostio-Eriophoretum tristis (F15, F16; QS = 50 and 52 respectively). A typical relevé (Tables 10-11, anal. 9) is from fairly dry ground, 60 m a.s.l. In one relevé only: *Carex glacialis, C. subspathacea, Tofieldia coccinea, Aulacomnium palustre, Orthothecium strictum, Hypnum revolutum, Drepanocladus revolvens, Mnium thomsonii, Cirriphyllum cirrhosum, Amphidium lapponicum, Amblyodon dealbatus.*

Typical subtype (E14.1)
Four relevés (75-300 m a.s.l.) from Ymer Ø differ in the overall dominating *Betula nana*. In the field notes the relevés are termed "striped *Betula* vegetations". pH of the topsoil 5.9-6.7, of the underground 6.5-7.3 (4 anal.). In one relevé only: *Draba glabella, Pyrola grandiflora, Ranunculus auricomus* var. *glabrata* (the only occurrence in the analyses from the 73°N area, Sørensen 1933 p. 53), *Aneura pinguis, Tortula ruralis.*

5.2. Middle-arctic grassland and fens

According to Sørensen (s.a.) "Group F" includes "fen-types on less basic ground (than Group E) characterized first of all by *Arctagrostis latifolia*. They seem not to be paralleled in Scandinavia and may possibly be described as a special, high arctic alliance, Arctagrostideon latifoliae". In the 73°N area the group includes three, geographically separated units. "Group F", supplemented by 21 Zackenberg relevés, are here considered representatives of ass. Arctagrostio-Eriophoretum tristis.

Because of the often hummocky character of the vegetations the separation in different units is somewhat problematic. In extreme cases the top of the up to ½ m high hummocks may harbour mossy or even dry heaths types, sometimes with a 10-20 cm belt of snowbed vegetations on the sides, and on the often minerogenous, more or less wet ground in between, fen vegetations are seen.

Quite many mosses were only registered in this association: *Calliergon giganteum, C. sarmentosum, Cinclidium subrotundum, Drepanocladus badius, Hypnum pratense, Lophozia rutheana, Meesia triquetra, Philonotis tomentella, Tayloria lingulata*, and apart from only one relevé outside, *Drepanocladus revolvens.* Some further species, occurring in only one unit, are mentioned below.

Arctagrostis latifolia is most frequent on soil with pH 5-6, whereas most other species occurring in all groups (15-20) of this association seem fairly pH indifferent (Figs 17-18). Conductivity is mostly fairly high.

5.2.1. Middle-arctic grassland and hummocky mires on poor/acid soil

Ass. Arctagrostio-Eriophoretum tristis Daniëls & Fredskild ass. nov. (F15-20, Z7-11) Preferential character-species are *Arctagrostis latifolia* and *Eriophorum triste*, nomenclatoric type-relevé: Table 12, anal. 89. The association, represented by 83 relevés (Tables 12-17), is divided into three subassociations: typicum, betuletosum, and juncetosum castanei.

Subass. typicum subass. nov. (Z7)
Grasslands, often heavily grazed by muskoxen, cover fairly large lowland areas at Zackenberg. By mid-August the surface, which is almost covered by mosses, is dry, reflected in the total absence of *Eriophorum scheuchzeri*. The ground is level or only slightly (<5°) sloping. Characteristic species are *Arctagrostis latifolia, Carex bigelowii, Juncus biglumis, J. castaneus. Carex capillaris* is frequent. The type relevé (Table 12, anal. 89) is from level ground c. 15 m a.s.l., on slightly higher ground gra-

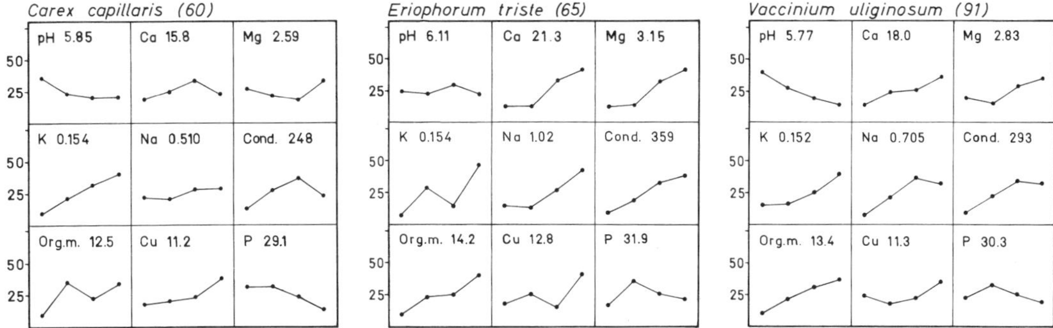

Fig. 17. Soil characteristica for three species common in ass. Arctagrostio-Eriophoretum tristis.

dually changing to a drier *Carex bigelowii-C. capillaris* grassland rich in lichens and with some *Carex rupestris* and *Hierochloë alpina*, but with only few *Eriophorum triste* and, especially, *Arctagrostis latifolia*. In one relevé only: *Luzula wahlenbergii*, which in Greenland is very rare, having its total distribution area between 74° and 76°N in East Greenland (Bay 1992). The six relevés are given in Fredskild & Bay (1993, Table 7, anal. 88-90, 92-94).

Further north, grasslands are common in the inland between 75° and 77½°N. On poor soil the dominating species are *Carex bigelowii, Arctagrostis latifolia, Carex misandra*, and *Eriophorum triste*, indicating their affinity to this association. Locally, on rich soil, they are accompanied by *Carex atrofusca, Kobresia simpliciuscula, Juncus triglumis, Eriophorum callitrix*, and *Pedicularis flammea*. In this part of Northeast Greenland Carex stans is only found in fens, often with *Eriophorum scheuchzeri*. North of 77½° grasslands are fewer, and *Carex stans* is also growing in these, replacing the southern *Carex bigelowii*, which has its northernmost occurrence at 78°30'N.

Variant of *Koenigia islandica* var. nov. (F18)

Five of six lowland, outer coast relevés at Hold with Hope are neighbouring "Flag vegetations", reflected in the similarity index (QS = 50) with Koenigio-Saginetum intermediae (C7), almost the same (QS = 49) as with the two following variants (F19-20). Differential species are *Koenigia islandica* and *Juncus triglumis*. A thick moss carpet is found in all relevés. Dwarf-shrubs make up only 7% of Summa F, the lowest of all groups in the association. A typical relevé (Table 12, anal. 48) is termed "*Carex saxatilis* swamp in connection with Flag". In one relevé only: *Deschampsia brevifolia*, and in the typical relevé: *Ranunculus hyperboreus* (10%).

Variant of *Vaccinium microphyllum* var. nov. (F19)

Three lowland relevés at Hold with Hope are characterized by *Vaccinium uliginosum* ssp. *microphyllum, Salix arctica*, and even *Dryas octopetala*, growing with *Arctagrostis latifolia, Carex bigelowii*, and *Eriophorum triste* in a thick moss carpet, in the two relevés dominated by *Sphagnum* and *Aulacomnium turgidum*. Judging from the field notes, the relevés are not hummocky. As to QS they are related to F18 (QS = 49) and F20 (QS = 52), and to some snowbed vegetations at Hold with Hope (D10, QS = 51). Besides, they are related to the hummocky mires in the inland on Kap Hedlund (F15, QS = 49), to which they seem to form an outer coast parallel. In the inland mires *Betula nana* and *Cassiope tetragona* are replacing *Vaccinium uliginosum*.

Variant of *Saxifraga oppositifolia* var. nov. (F20.1)

Five middle altitude relevés on Traill Ø, the four at 500-550 m a.s.l., are characterized by a high diversity of forbs, gramino-

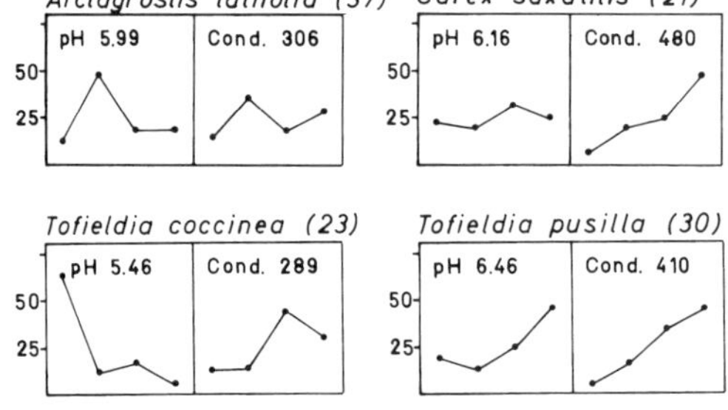

Fig. 18. pH and conductivity for five species common in ass. Arctagrostio-Eriophoretum tristis.

Subassociation/variant	F18			Z7			F19			F20.1	20.2	
Area	HH			Z			HH			T		
Elevation, m	0-100			10-75			0-100			30-550		
No. of relevés			6			6				5	3	8
F% (F) or average F% (av.)		av.		F	av.		F	F	F	av.	av.	
Analysis no.	48			89			12	50	30			
Polygonum viviparum	90	98	6	90	92	6	100	100	100	88	90	8
Salix arctica	50	72	5	70	67	5	100	100	80	96	97	8
Eriophorum triste	80	80	5	90	63	5	100	100	100	54	95	7
Arctagrostis latifolia	100	93	6	100	70	6	100	20	80	48	72	6
Carex bigelowii	50	72	6	90	78	6	100	100	40	78		4
Stellaria longipes s.l.	10	48	6		2	2	20	10	10	28	3	4
Luzula arctica		37	5	10	3	2	40			53	48	7
Equisetum arvense	60	37	4	20	20	3			90	14	60	5
Pedicularis hirsuta	10	33	4				40	70	10	35	10	5
Vaccinium microphyllum		3	2	+	8	5	100	100	100	1	2	2
Carex saxatilis	100	75	5	100	53	4			80	8		1
Poa arctica		5	1		22	4		20		29	2	5
Luzula confusa		12	3		6	2		10		42	10	4
Equisetum variegatum		15	1				100	10		8	48	5
Silene acaulis		5	2						10	12		3
Koenigia islandica	60	53	6									
Carex maritima		27	3									
Eriophorum scheuchzeri	90	32	2									
Melandrium apetalum		5	2									
Carex atrofusca		15	1									
Carex parallela		10	1									
Juncus castaneus	80	40	5	80	42	5						
Alopecurus alpinus	40	15	2	10	3	3						
Juncus biglumis	50	65	6		53	5				35	10	6
Carex misandra		33	4		+	1				91	28	8
Saxifraga cernua		20	3		7	2				36	2	6
Carex rupestris		2	1		23	4				25		4
Potentilla hyparctica		13	3		10	2				23		4
Saxifraga foliolosa	30	20	5		2	1				10	10	2
Cardamine bellidifolia		15	3							18	13	6
Draba lactea		13	3							20	7	7
Juncus triglumis	100	72	5							2		1
Saxifraga tenuis	10	7	3							7	2	3
Ranunculus sulphureus		7	2							4		3
Sagina intermedia		3	2							1		1
Carex capillaris				20	32	5						
Hierochloë alpina					18	2						
Carex norvegica					3	2						
Pedicularis flammea					3	2						
Dryas octopetala/sp.					2	2	70	50	40	56	55	8
Festuca brachyphylla/s.l.					8	3		10		16		3
Saxifraga oppositifolia										85	53	8
Cassiope tetragona										17	100	4
Oxyria digyna										5	2	2
Taraxacum arcticum										10		4
Minuartia rubella										2		2
Huperzia selago											15	2
No. of species, range	17-25			12-18			11-13			21-24	14-21	
No. of species, average	21			15			12			23	17	
Sum of F%, range	1010-1340			560-900			700-870			835-1340	625-985	
Sum of F%, average	1155			688			770			1102	833	

Table 12. Ass. Arctagrostio-Eriophoretum tristis subass. typicum, phanerogams.

Community	20.1	20.2	
No. of relevés	5	2	7
Average F% (av.)	av.	av.	
Ditrichum flexicaule	86	60	7
Distichium capillaceum	64	30	7
Tortella fragilis	62	10	6
Isopterygium pulchellum	42	30	6
Myurella julacea	46	5	6
Polytrichum strictum	46	20	5
Fissidens osmundoides	24	15	5
Aulacomnium turgidum	18	70	4
Cladonia pyxidata	20	15	4
Pohlia cruda	4	35	3
Campylium stellatum	16	10	3
Philonotis tomentella	16	5	3
Cyrtomnium hymenophylloides	12	5	3
Drepanocladus uncinatus	2	35	2
Hypnum bambergeri	20	10	2
Tomenthypnum nitens	2	20	2
Orthothecium chryseum	2	5	2
Bryoerythrophyll. recurvirostre	20		4
Encalypta alpina	8		3
Scorpidium turgescens	20		2
Nostoc sp.	16		2
Meesia uliginosa	8		2
Myurella tenerrima	4		2
Dicranum spadiceum		55	2
Blepharostoma trichophylla		20	2
No. of species, range	13-17	17-21	
No. of species, average	16	19	
Summa F%, range	600-770	630-650	
Summa F%, average	684	640	

Table 13. Ass. Arctagrostio-Eriophoretum tristis (F15), cryptogams. Var. of *Saxifraga oppositifolia* (F20.1) and subvar. of *Cassiope tetragona* (F20.2).

ids, and mosses. The moss carpet is interrupted by dark spots where organic crust cover the sandy-stony underground, pH of which is 5.3-6.1 (5 anal.). The many, small forbs and the organic crust indicate a mosaic vegetation, being fairly common further north in East Greenland on slopes below snowdrifts. In one relevé only: *Campanula uniflora, Chamaenerion latifolium, Draba adamsii, Papaver radicatum, Poa pratensis* var. *colpodea, Ranunculus glacialis, Amphidium lapponicum, Barbula asperifolia, Calliergon sarmentosum, Drepanocladus intermedius, D. revolvens, Timmia austriaca, Cladonia bellidiflora.* Cryptogam analyses in Table 13.

Subvariant of *Cassiope tetragona* (F20.2)
Three relevés from lower altitude on Traill Ø were termed "*Cassiope* myr". *Dicranum spadiceum*, common in poor *Cassiope* heaths, is very frequent in the two relevés

	F15									
Variant/community	F15.1		F15.2			Z8			F16	
Area	H		H			Z			H	
Elevation, m	30-40		30-40			10-50			30-450	
No. of relevés		3		4	7					5
F% (F) or average F% (av.)	F	av.	F	av.		F	F	F	av.	
Analysis no.	69		267			40	41	38		
Eriophorum triste	100	98	100	100	7	100	40	100	76	5
Vaccinium microphyllum	100	100	100	81	7			10	67	5
Polygonum viviparum	85	67	100	95	7	100	100	80	89	5
Salix arctica	95	67	75	68	7	100	100	100	60	5
Betula nana	55	45	25	66	7				21	3
Equisetum arvense	100	100	90	70	6	40	50	70	43	3
Dryas octopetala/sp.		20	80	65	5		30	10	33	4
Carex capillaris		2		29	5	30	20		21	4
Arctagrostis latifolia	45	92	40	53	7	40		90	1	1
Carex rupestris	10	3	75	68	5	10			30	2
Carex bigelowii		33	30	33	3		40		32	4
Cassiope tetragona	65	55	5	30	5			30	2	2
Luzula confusa		20		1	3	+			2	2
Pedicularis hirsuta		3	10	9	3			10	1	1
Poa arctica		30			2	90	10		4	1
Eriophorum scheuchzeri	5	2			1	20		40	30	2
Cardamine bellidifolia	5	7	5	13	5					
Pyrola grandiflora		28		28	3					
Pedicularis lapponica	10	7		1	3					
Draba lactea		5		1	2					
Luzula arctica	5	17	5	16	6		60	50		
Tofieldia coccinea	15	18		5	5				16	4
Tofieldia pusilla	5	2		3	2				49	4
Saxifraga oppositifolia		5	15	25	4				5	2
Silene acaulis		8	10	18	5				6	1
Carex parallela		3		46	4				6	1
Empetrum hermaphroditum	25	35		1	4				2	1
Juncus biglumis			40	18	3	90	60	50	2	2
Kobresia myosuroides				10	1	10			24	2
Equisetum variegatum				15	1		100			
Saxifraga cernua				5	1	30	40			
Carex misandra			20	21	4				46	5
Pedicularis flammea				8	3				38	5
Carex saxatilis				3	2				51	4
Juncus castaneus			90	24	2				30	4
Eriophorum callitrix			20	6	2				25	4
Juncus triglumis			10	3	1				32	3
Carex scirpoidea				16	1				15	2
Saxifraga foliolosa			15	4	1				1	1
Alopecurus alpinus						60	70	40		
Festuca brachyphylla						30	80	10		
Saxifraga hirculus						10	20			
Rhododendron lapponicum									57	5
Carex atrofusca									40	4
Carex rariflora									57	3
Kobresia simpliciuscula									25	3
No. of species, range	12-23		23-26			14-18			21-24	
No. of species, average	17		25			16			22	
Summa F%, range	685-1155		990-1205			690-900			820-1210	
Summa F%, average	872		1069			783			1067	

Table 14. Ass. Arctagrostio-Eriophoretum tristis subass. betuletosum nanae (F15), phanerogams. Further, subass. typicum var. of *Alopecurus alpinus* (Z8) and *Eriophorum triste-Rhododendron lapponicum* comm. of subass. eriophoretosum scheuchzeri (F16).

with cryptogam analyses (Table 13). pH is 5.2-5.5 (2 anal.). In one relevé only: *Anastrophyllum minutum, Aulacomnium palustre, Bartramia ityphylla, Brachythecium salebrosum, Drepanocladus badius, Hylocomium splendens, Hypnum callichro-*

Variant/community	F15					F16	
	F15.1		F15.2			F16	
No. of relevés		*3*		*4*	*7*		*5*
F% (E) or average F% (av.)	F	av.	F	av.		av.	
Analysis no.	69		267				
Ditrichum flexicaule		23	80	40	*4*	30	*5*
Blepharostoma trichophyllum	10	20	30	30	*6*	34	*3*
Aulacomnium turgidum	90	47	90	87	*6*	17	*3*
Campylium stellatum		13	70	40	*5*	16	*3*
Oncophorus wahlenbergii		3	20	13	*4*	23	*4*
Distichium capillaceum		7	70	45	*5*	9	*2*
Drepanocladus badius	90	27		10	*3*	44	*3*
Isopterygium pulchellum		3		10	*3*	23	*3*
Pohlia cruda		23		27	*3*	9	*1*
Sphenolobus minutus	10	3		10	*2*	4	*2*
Aulacomnium palustre	60	40			*2*	16	*3*
Polytrichum strictum	50	20			*2*	1	*1*
Drepanocladus revolvens			10	10	*2*	47	*3*
Meesia uliginosa				3	*1*	23	*3*
Myurella julacea			50	20	*2*	4	*2*
Polytrichum alpinum				8	*2*	9	*2*
Cladonia pyxidata				13	*1*	13	*2*
Odontoschisma macounii				7	*1*	24	*2*
Tortella fragilis			20	20	*2*	1	*1*
Tomenthypnum nitens	90	63	50	73	*5*		
Pohlia nutans	30	17		27	*3*		
Drepanocladus uncinatus		23			*2*		
Philonotis tomentella			10	10	*2*		
Calliergon trifarium						33	*3*
Cinclidium subrotundum						44	*2*
Calliergon sarmentosum						27	*2*
Leiocolea rutheana						7	*2*
Fissidens osmundoides						3	*2*
No. of species, range	6-17		11-16			7-18	
No. of species, average	11		14			13	
Summa F%, range	170-640		410-560			370-630	
Summa F%, average	407		508			513	

Table 15. Ass. Arctagrostio-Eriophoretum tristis subass. betuletosum nanae (F15) and *Eriophorum triste-Rhododendron lapponicum* comm. of subass. eriophoretosum scheuchzeri (F16), cryptogams.

um, H. pratense, Odontoschisma macounii, Oncophorus wahlenbergii, Pohlia nutans, Polytrichastrum alpinum.

Besides being related to F18 and F19 (QS = 49 and 52), F20 is related to Saxifrago-Kobresietum simpliciusculae (E13, QS = 53), mainly because of the the floristic composition of the *Carex misandra-Luzula arctica* community (F20.1), and to the *Cassiope-Vaccinium uliginosum* heaths (J30, QS = 50) and *Dryas octopetala-Carex rupestris-Carex bigelowii* heaths (J31, QS = 50) of Dryadion integrifoliae.

Variant of *Alopecurus alpinus* var. nov. (Z8)

In the lowland at Zackenberg a common vegetation type, in Fredskild & Bay (1993) termed "snowpatch-heaths", is dominated by *Salix arctica* and *Eriophorum triste*. The ground is hummocky or patterned. In the latter case this vegetation is found on ½-1½ m wide, stabilized polygons, elevated c. 10 cm above the intervening, usually less than ½ m wide "channels", covered by *Eriophorum scheuchzeri* fens with e.g. *Alopecurus alpinus* and *Poa arctica*, as is the case with a typical relevé (Table 14, anal. 40). The ground, 27 m a.s.l., is level or less than 5° sloping. Mosses cover all of the ground in both vegetations. The analyses are from the center of 10 randomly selected polygons. In one relevé only: *Stellaria longipes* s.l. (70%), *Ranunculus sulphureus* (10%), and *Carex norvegica* (+), all in anal. 41, and, in the typical relevé, *Kobresia myosuroides* (10%).

Subass. betuletosum nanae subass. nov. (F15)

Typical variant var. nov. (F15.1)

Three lowland relevés (30-40 m a.s.l.) are from solifluction lobes on N-facing slopes with a fairly long lasting snow cover on Kap Hedlund. A typical relevé (Tables 14-15, anal. 69) is upslope replaced by *Cassiope tetragona* heaths (G21.1 and G21.3), downslope by another of the three relevés of the variant, a very hummocky *Carex bigelowii-Vaccinium uliginosum* mire. pH of the topsoil is 4.5-5.4, of the underground 4.9-5.2 (2 anal.). Differential taxa against subass. eriophoretosum scheuchzeri (Z9) and Arctagrostio-Eriophoretum tristis (Z7) are *Betula nana, Tofieldia coccinea, T. pusilla, Empetrum hermaphroditum, Pyrola grandiflora, Pedicularis lapponica,* and *Oncophorus wahlenbergii* (Table 15). 34% of the sum of F% are graminoids, 37% dwarfshrubs. The acid soil with fairly low conductivity is confirmed by the frequent occurrence of species like *Vaccinium uliginosum, Empetrum hermaphroditum, Tofieldia coccinea* (Figs 17-19). In one relevé only: *Hypnum pratense* (anal. 69, 30%).

Variant of *Carex rupestris* var. nov. (F15.2)

Four lowland relevés (30-40 m a.s.l.) from solifluction slopes on Kap Hedlund differ in the frequent *Carex rupestris, C. capillaris, C. misandra*, and *Dryas octopetala*. They are less hummocky. Graminoids are 42%, dwarfshrubs 29% of Summa F. pH is 5.1-5.9 (2 anal.). A typical relevé (Tables 14-15, anal. 267), by Sørensen termed "*Eriophorum* Myr", is almost without hummocks. In one relevé only: *Cerastium alpinum, Draba glabella, Rumex acetosella, Orthothecium intricatum, Scorpidium turgescens*, and in the type relevé *Carex norvegica* (15%) and *Myurella tenerrima* (10%). In the 73°N area *Carex norvegica* only appears here and in another relevé (J25) from Kap Hedlund.

5.2.2. Middle-arctic fens

Subass. eriophoretosum scheuchzeri subass. nov. (Z9-11, F17)

At Zackenberg true fens, i.e. permanently wet vegetations dominated by graminoids and with mosses covering all the ground, are only found in the lowland on a large plain, formed by marine, glaciofluvial, fluvial, and deltaic deposits, sedimentated between old, eroded moraines. With the exception of narrow rims at the edge of some ponds, *Salix arctica* and *Polygonum viviparum* are constant. In the most wet fens the growth of *Salix arctica* resembles that of the low-arctic *S. arctophila*, with creeping, rooting branches. Contrary to *Eriophorum scheuchzeri, E. triste* is here mainly sterile. The type relevé, 11 m a.s.l.,

Subass./variant	Z9			F17.1					F17.2		Z10			Z11		
Area	Z			EY			Z		T		Z			Z		
Elevation, m	10-55			20-200			10-30		225		25-50			25-40		
No. of relevés			*4*													
F% (F) or average F% (av.)	F	av.		F	F	F	F	F	F	F	F	F	F	F	F	F
Analysis no.	75			326	327	121	76	84	235	236	72	73	77	80	78	81
Salix arctica	70	75	*4*	55	40	95	80	30	100	100	40	90	90	40	60	80
Polygonum viviparum	60	83	*4*	15	5	100	100	30	100	100	100	90	90	10	60	70
Arctagrostis latifolia	100	100	*4*	25	95		100	30	100	60	90	100	100	80	80	100
Eriophorum scheuchzeri	100	65	*4*	95	100	20	60	40		10	100	70	100	80	60	90
Eriophorum triste	60	60	*3*	70	5	100	60	40	100	100		20	50	40	60	90
Equisetum arvense	30	53	*4*			55	30		90	90			10	10	80	90
Carex saxatilis	100	58	*4*	100	100	100	70	70	100	90			100			
Equisetum variegatum		10	*1*	100	100	100	20		100	90						40
Juncus biglumis	30	45	*4*			20	+	40				10	10	30		
Juncus castaneus	30	18	*4*	75				10						10		
Carex pseudolagopina	10	3	*1*		35		100									
Pedicularis flammea			+	1		15										
Dryas octopetala/sp.		13	*2*					30	10				10			10
Carex bigelowii	40	60	*3*					30			10					
Saxifraga foliolosa		3	*1*				10				+	10	10			
Carex misandra		3	*1*					10								
Poa arctica		8	*1*											100		
Dupontia psilosantha		3	*1*								100	100	50			
Saxifraga cernua		28	*2*								30	40				
Luzula wahlenbergii	+	3	*2*										20			
Alopecurus alpinus		8	*1*											90	100	20
Eriophorum callitrix				50	10	40	90	+								
Juncus triglumis				15		30		20								
Kobresia simpliciuscula				5		5		+								
Carex rariflora				100				100								
Carex atrofusca				20				20								
Carex parallela						100			100	100						
Vaccinium microphyllum								20	100							
Pedicularis hirsuta									20	10		+				
Draba lactea									10							+
No. of species, range	9-15			13	9	15	12	17	11	12	8	14	11	11	8	10
No. of species, average	13															
Summa F%. range	580-780			490-810			520-720		760-930		470-630			490-590		
Summa F, average	695			675			620		845		560			530		

Table 16. Ass. Arctagrostio-Eriophoretum tristis subass. eriophoretosum scheuchzeri, phanerogams.

Variant	17.1			17.2	
Analysis no.	326	327	121	235	236
Drepanocladus brevifolius	100	100	30		10
Campylium stellatum		50	80	30	10
Meesia uliginosa	80	40		50	10
Ditrichum flexicaule		30	40	60	
Calliergon giganteum	90	40			30
Cinclidium arcticum		70		30	60
Drepanocladus badius		20		50	10
Distichium capillaceum	20		30		10
Tortella fragilis			10	10	
Calliergon trifarium			10		10
Aneura pinguis	10				10
Catoscopium nigritum	70	30	30		
Scorpidium turgescens		30	50		
Cinclidium subrotundum	100				
Timmia norvegica		20			
Nostoc sp.			90		
Hypnum bambergeri			30		
Aulacomnium palustre				10	80
Tomenthypnum nitens				60	30
Meesia triquetra				40	50
Leiocolea rutheana				30	10
Drepanocladus revolvens				100	
Blepharostoma trichyphylla				40	
Tayloria lingulata				20	
Polytrichum alpinum				20	
Orthothecium chryseum				10	
Dicranum angustum				30	
Barbilophozia kunzeana				10	
Blindia acuta				10	
Drepanocladus intermedius					100
No. of species	10	12	12	19	15
Summa F%	540	520	530	670	480

Table 17. Ass. Arctagrostio-Eriophoretum tristis subass. eriophoretosum scheuchzeri, cryptogams.

is only slightly hummocky, with Salix on the hummocks (Table 16, anal. 75). In mid-August the ground was very humid, not wet. However, in the neighbouring, more hummocky part of the fen, still some clear water was visible between the hummocks. From here, another of the relevés (Fredskild & Bay 1993, anal. 85) of the subass. originates. The two remaining relevés are anal. 79 and 86 (Fredskild & Bay l.c.). Differential species are *Juncus castaneus, Carex saxatilis*, and *Alopecurus alpinus. Carex saxatilis*, growing on soils with the highest average conductivity in the 73°N area (Table 2), indicates rich soils for the subassociation. *Eriophorum scheuchzeri* is a good differential species. In one relevé only (the type relevé): *Carex capillaris* (10%).

Variant of *Eriophorum callitrix* var. nov. (F17.1)

Eriophorum callitrix is a selective species in fens on the more rich soils at Zackenberg as well as in the 73°N area. A typical relevé (Tables 16-17, anal. 326) is from a "*Carex saxatilis* swamp, ½ m deep" in a river delta, 20 m a.s.l. on Ymer Ø. Presumably the depth given refers to living material plus peat. This vegetation gradually changes to another of the relevés, a "*Carex saxatilis-Arctagrostis* swamp, ½ m deep", with *Carex pseudolagopina* but without *C. rariflora* and *Juncus castaneus*. pH is 6.8-7.2. In one relevé only: *Carex bicolor* and *C. scirpoidea* The moss analyses are given in Table 17, which includes the only occurrence in the 73°N area of *Timmia norvegica*. Two relevés from Zackenberg (Fredskild & Bay 1993, anal. 76 and 84) belong to the variant.

Variant of *Carex parallela* var. nov. (F17.2)

Two relevés from hummocky mires, 225 m a.s.l. on Traill Ø are termed "*Vaccinium* mire" and "Pure *Salix* mire", respectively, in the field notes of Sørensen. The moss samples (Table 17) were taken between the hummocks. These are the only relevés with *Barbilophozia kunzeana, Blindia acuta*, and *Dicranum angustum* in the 73°N area. pH in the *Vaccinium* mire is 6.0. In one relevé only: *Melandrium apetalum* (anal. 236, 10%).

Variant of *Dupontia psilosantha* var. nov. (Z10)

The most widespread fens in the Zackenberg lowland are either typical eriophoretosum scheuchzeri fens or representatives of the variant of *Dupontia psilosantha*. A typical relevé of this (Table 16, anal. 72) is from a uniform, not hummocky 30 m wide belt between a moist heath and a hummocky fen on slightly higher ground. Differential species is *Dupontia psilosantha*. The surface at anal. 73 was fairly dry early in August. This, as well as anal. 77 had low hummocks. In one relevé (anal. 73) only: *Festuca brachyphylla* and *Potentilla hyparctica*, both 10%.

Variant of *Alopecurus* alpinus var. nov. (Z11)

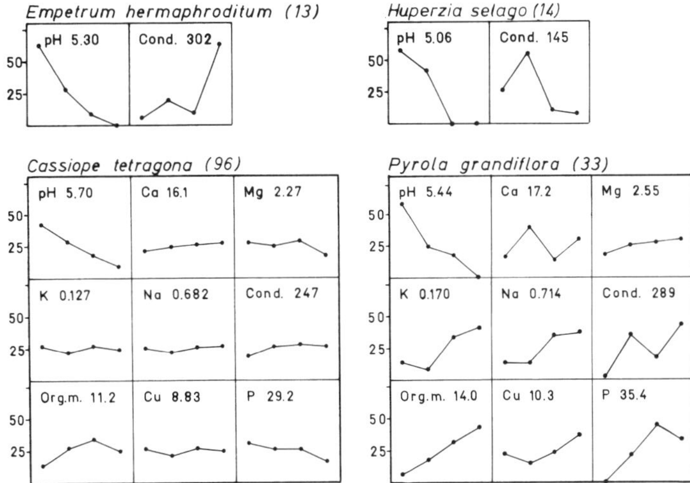

Fig. 19. Soil characteristica for four species common in ass. Saliceto-Cassiopetum tetragonae.

Locally, *Alopecurus alpinus* is common in some fens or fen-like vegetations in the Zackenberg lowland, represented here by three relevés. Anal. 80 (Table 16) is from a "channel" between 1-1.5 m wide, slightly elevated polygons with the typical relevé of the var. of *Alopecurus alpinus* (Z8). Anal. 78 is from a similar "channel", with a *Salix arctica-Polygonum viviparum-Alopecurus alpinus-Eriophorum triste-Equisetum arvense-E. variegatum* vegetation on the polygons. Anal. 81 is from a slightly hummocky vegetation along a brooklet, with some water-deposited clay on the moss carpet. In one relevé only: *Luzula arctica* (anal. 78, 10%) and *Stellaria longipes* (anal. 80, +).

Eriophorum triste-Rhododendron lapponicum community (F16)

Two lowland relevés from the same slope on Kap Hedlund as two of the relevés of F15.1, and three from higher level (425-475m) are characterized by the frequent *Rhododendron lapponicum* (Tables 14-15). Two of the latter relevés are from stony lake margins, the third from a slope below a *Ranunculus hyperboreus-Phippsia algida* snowbed. pH of the topsoil is 4.5-5.7 (5 anal.), of the deeper 4.6-5.0 (2 anal.) which confirms the affinity of the community to the hummocky mires (QS = 51 to F15). However, QS is even higher (52) to as well E14 (Saxifrago-Kobresietum simpliciusculae var. of *Betula nana* and subtype of *Tofieldia pusilla*) as J25 (*Tofieldia coccinea-Carex bigelowii* comm. of Rhododendro-Vaccinietum microphylli). The main difference against E14 is the absence in F16 of *Saxifraga aizoides, Equisetum variegatum, Braya purpurascens*, all on richer soil, of *Saxifraga nathorstii*, never found in the inland, and of *Pedicularis lapponica*, and the absence in E14 of *Empetrum hermaphroditum* and *Eriophorum scheuchzeri*. Against J25 the main difference is the absence of *Carex nardina* and *Pedicularis lapponica*, and the more sparse *Kobresia myosuroides* in F16, and the absence of *Carex saxatilis, C. atrofusca, C. rariflora*, and *Eriophorum callitrix* in J25, illustrating the difference between the drier heaths and the intermediate "heath mire".

The community seems related to Rhododendro-Vaccinietum microphylli Daniëls 1982.

Arctagrostis latifolia is found in only one of 20 circles in one of the five relevés of F16. On the contrary it is found, often dominating, in 46 of 50 relevés in the other units of Arctagrostio-Eriophoretum tristis. Considering the problems in classifying hummocky communities phytosociologically the question where to place the *Eriophorum triste-Rhododendron lapponicum* community is left open.

In one relevé only: *Euphrasia frigida, Hierochloë alpina, Saxifraga aizoides, Amphidium lapponicum, Anthelia juratzkana, Cetrariella delisei, Cyrtomnium hymenophylloides, Orthothesium chryseum, Tayloria lingulata*, and the only occurrence in the 73°N area of *Anoectangium aestivum*.

5.3. Middle-arctic dwarfshrub and graminoid heaths on mesic to dry, mainly circumneutral -basic soil

Ass. Rhododendro-Vaccinietum microphylli Daniëls 1982 (J24-25)
53 relevés from dwarfshrub and graminoid heaths are provisionally grouped under Rhododendro-Vaccinietum microphyllae, described as: *Betula nana-Tofieldia pusilla* community and *Tofieldia coccinea-Carex bigelowii* community.

Betula nana-Tofieldia pusilla community (J24)
39 relevés from Ella Ø and Ymer Ø were by Sørensen divided into two groups: 24.1 and 24.2, the latter further subdivided into four groups. As seen from the dendrogram (Fig. 3) the separation of these four takes place at a very high level (QS = 66), also reflected in Table 18, in which the subdivision is retained. *Tofieldia pusilla* is one of the characteristic species, otherwise being fairly common only in the closely related *Eriophorum triste-Rhododendron lapponicum* community (F16). Other characteristic species, yet with a wider phytosociological amplitude, are *Betula nana, Carex scirpoidea, Pedicularis lapponica, Cassiope tetragona, Carex nardina, C. misandra,* and *Orthothecium intricatum.*

Subtype of *Carex scirpoidea* (J24.2b)
The 13 relevés are all from the lowland below 140 m a.s.l. Between the chamaephytes, mainly *Dryas octopetala, Cassiope tetragona, Vaccinium uliginosum, Betula nana,* and *Rhododendron lapponicum,* many forbs, graminoids, and mosses are seen, making this subtype one of the most species rich in the 73°N area. *Rhododendron lapponicum* is constant, yet with a very varying F%. The ground is level or only slightly sloping, locally hummocky, often with a topmost, up to 15 cm thick peaty layer overlying gravel. pH of the topsoil is 5.3-7.1 (9 anal.), of the subsoil 6.3-7.1 (5 anal.). A typical relevé (Tables 18-19, anal. 31) is from level ground, 80 m a.s.l. pH of the 10 cm thick topsoil 6.1, below this 6.5. As indicated by the occurrence of e.g. *Pedicularis flammea, Tofieldia pusilla,* and *Carex scirpoidea* the soil does not dry out during summer. *Betula nana* seems pH indifferent yet preferring soil with a fairly high conductivity (Fig. 21). A dry *Dryas-Rhododendron* heath at Segelsällskapets Fjord 35 km WNW of Mestersvig is a representative of the subassociation (Elkington 1965, anal. 11). Regional differential taxa are *Carex misandra, Salix arctica,* and *Equisetum variegatum.*

Subtype of *Kobresia myosuroides* (J24.2a)
The eight relevés are fairly dry, indicated by *Kobresia myosuroides* (F% = 43) and few or none *Carex scirpoidea, Tofieldia pusilla,* and *Saxifraga aizoides.* pH of topsoil 6.1-7.5 (6 anal.), subsoil 6.6-6.8 (2 anal.).

Subtype of *Equisetum variegatum* (J24.2c-d)
The six relevés of J24.2c are from slightly sloping ground, in some cases forming a belt with a longer snow cover below vegetations of 24.2d. pH of the topsoil 5.7-6.8, subsoil 6.7-7.2 (6 anal).

In the six relevés of J24.2d *Cassiope tetragona* and *Vaccinium uliginosum* are mostly missing. pH of topsoil 6.6-7.2 (6 anal.), subsoil 7.0-7.3 (3 anal.). The vegetation seems to be snow covered too long for *Cassiope* to grow here. In one relevé only: *Lesquerella arctica.*

Subtype of *Empetrum hermaphroditum* (J24.1)
6 relevés from slopes with a southern exposure are charcterized by dwarfshrubs (av. sum of F%=60), mainly *Vaccinium uliginosum, Cassiope tetragona,* and *Empetrum hermaphroditum,* and with frequent *Pyrola grandiflora. Empetrum hermaphroditum* is characteristic, otherwise occurring in only two of the remaining 47 relevés of the association. In the 73°N area *Poa pratensis* ssp. *alpigena* attains the highest frequency here and in the subtype of *Euphrasia frigida* (J25.3). *Dryas octopetala,*

	J24													J25							
Subass./variant/community	J24.1			J24.2.a		J24.2.b			J24.2.c		J24.2.d			J25.1			J25.2		J25.3		
Area	E			E		E			E		E			H			T		E		
Elevation, m	80-400			30-175		20-140			30-350		175-500			50-450			60-550		200		
No. of relevés			6		8			13		6		6	39			6		4		4	14
F% (F) or average F% (av.)	F	av.		av.		F	av.		av.		av.			F	av.		av.		av.		
Analysis no.	33					31								45							
Dryas octopetala	75	65	*6*	87	*8*	90	93	*13*	97	*6*	96	*6*	*39*	35	59	*6*	99	*4*	100	*4*	*14*
Carex rupestris	90	42	*6*	87	*8*	75	95	*13*	46	*6*	71	*5*	*38*	100	99	*6*	95	*4*	100	*4*	*14*
Polygonum viviparum	85	64	*6*	63	*7*	50	60	*12*	93	*6*	93	*6*	*37*	100	87	*6*	96	*4*	100	*4*	*14*
Silene acaulis	30	22	*6*	34	*8*	55	61	*13*	45	*6*	35	*6*	*39*		23	*4*	9	*3*	79	*4*	*11*
Saxifraga oppositifolia	45	17	*4*	80	*8*	100	97	*13*	93	*6*	93	*6*	*37*		60	*4*	49	*3*	99	*4*	*11*
Vaccinium microphyllum	100	100	*6*	54	*5*	100	72	*12*	86	*6*	4	*1*	*30*	100	48	*6*	99	*4*	55	*4*	*14*
Cassiope tetragona	100	88	*6*	81	*8*	100	97	*13*	97	*6*	2	*2*	*35*		3	*3*	29	*4*	4	*1*	*8*
Salix arctica	60	54	*6*	32	*6*	20	24	*9*	48	*6*	54	*6*	*33*	35	34	*5*	24	*2*	21	*3*	*10*
Betula nana	90	43	*5*	66	*7*	100	67	*11*	74	*6*	87	*6*	*35*	10	16	*3*			3	*1*	*4*
Carex misandra	5	7	*2*	36	*7*	15	55	*12*	39	*5*	18	*4*	*30*		8	*2*	34	*2*	85	*4*	*8*
Pedicularia flammea		2	*2*	8	*5*		35	*9*	36	*4*	22	*4*	*24*	20	23	*4*	34	*3*	81	*4*	*11*
Carex nardina		3	*1*	51	*7*	10	58	*11*	4	*3*	13	*4*	*26*		13	*1*	20	*2*	66	*4*	*7*
Kobresia myosuroides		1	*1*	43	*7*		17	*7*	4	*2*	8	*2*	*19*	100	58	*6*	35	*2*	69	*3*	*11*
Carex scirpoidea		44	*3*	3	*1*	90	29	*10*	83	*6*	89	*6*	*26*						28	*2*	*2*
Carex capillaris	5	2	*2*			40	22	*10*	4	*2*	1	*1*	*15*	60	76	*6*	70	*4*	53	*3*	*13*
Rhododendron lapponicum	15	3	*1*			15	36	*13*					*14*	60	63	*6*	75	*4*	75	*4*	*14*
Pedicularis lapponica	40	13	*3*	3	*3*	55	12	*7*	21	*4*	25	*4*	*21*	5	3	*2*					*2*
Tofieldia pusilla		13	*3*			100	52	*11*	43	*6*	5	*1*	*21*				3	*1*			*1*
Saxifraga aizoides						5	16	*4*	32	*4*	21	*4*	*12*						16	*3*	*3*
Tofieldia coccinea	10	2	*1*						1	*1*			*2*	15	48	*6*	33	*2*	73	*4*	*12*
Kobresia simpliciuscula				1	*1*		10	*3*	4	*1*	3	*3*	*8*				29	*2*	93	*4*	*6*
Arctostaphylos alpina		3	*2*	1	*1*	10	4	*4*	2	*1*	17	*1*	*9*						5	*2*	*2*
Pedicularis hirsuta	5	2	*2*	8	*2*				5	*2*	3	*3*	*9*		2	*1*	1	*1*			*2*
Equisetum arvense		38	*4*				10	*3*	3	*1*			*8*		16	*1*					*1*
Luzula confusa	10	2	*1*										*1*	10	11	*6*	13	*2*			*8*
Luzula arctica				1	*1*		14	*5*					*6*						15	*3*	*3*
Poa pratensis ssp. *alpigena*	15	27	*4*						3	*1*			*5*				3	*1*	20	*3*	*4*
Empetrum hermaphroditum	100	78	*6*	1	*1*								*7*		1	*1*					*1*
Armeria scabra ssp. *sibirica*		2	*1*	3	*1*		8	*2*	13	*1*			*5*						23	*3*	*3*
Eriophorum triste							5	*1*	3	*2*			*3*		16	*1*	15	*1*	30	*2*	*4*
Chamaenerion latifolium		13	*2*	1	*2*	5	B1	*1*	1	*1*			*5*		2	*1*	3	*1*			*2*
Juncus biglumis							1	*1*	2	*1*			*2*		3	*1*	16	*1*	51	*3*	*5*
Minuartia stricta							3	*2*	4	*3*			*5*						3	*1*	*1*
Carex parallela				1	*1*		1	*1*	7	*1*			*3*				23	*1*	23	*1*	*2*
Arctagrostis latifolia		10	*1*	1	*1*		1	*1*	1	*1*			*4*		13	*1*					*1*
Draba glabella		2	*2*				1	*1*					*3*	15	3	*2*					*2*
Juncus triglumis							1	*1*					*1*				23	*1*	40	*2*	*3*
Papaver radicatum		2	*1*										*1*	5	3	*3*					*3*
Eutrema edwardsii				1	*1*		1	*1*	2	*1*			*3*						1	*1*	*1*
Draba lactea									3	*1*			*1*		5	*1*					*1*
Melandrium triflorum/affine				1	*1*								*1*	5	1	*1*					*1*
Equisetum variegatum				10	*2*		17	*6*	60	*6*	15	*3*	*17*								
Pyrola grandiflora	95	53	*4*	13	*2*	5	7	*2*			8	*1*	*9*								
Saxifraga nathorstii							2	*1*	13	*2*	4	*2*	*5*								
Carex glacialis				1	*1*		7	*2*					*3*								
Braya purpurascens				1	*1*				4	*1*	1	*1*	*3*								
Arenaria pseudofrigida						5	1	*2*					*2*								
Carex bigelowii														85	43	*5*	54	*3*	25	*1*	*9*
Saxifraga cernua															2	*2*	6	*3*			*5*
Euphrasia frigida															4	*1*			35	*3*	*4*
Hierochloë alpina														50	33	*3*					*3*
Cardamine bellidifolia															5	*2*	1	*1*			*3*
Carex supina														5	17	*2*					*2*
Rumex acetosella														10	2	*1*	9	*1*			*2*
Woodsia glabella															1	*1*	3	*1*			*2*
No. of species, range	20	14-20		10-19		22	10-24		17-23		10-18			20	15-26		16-20		16-27		
No. of species, average		17		14			19		20		15				20		18		22		
Summa F%, range	980	705-980		540-990		1050	665-1450		970-1190		625-995			825	675-1205		660-1455		1000-1785		
Summa%, average		815		769			1092		1076		788				910		1011		1468		

Table 18. Ass. Rhododendro-Vaccinietum microphyllae, phanerogams.

Betula nana, *Carex misandra*, and others are markedly fewer than in the subtype of *Carex scirpoidea* (J24.2.b). pH of the topsoil 5.9-6.4 (4 anal.), subsoil 6.3-6.4 (3 anal.). A typical relevé (Tables 18-19, anal. 33) is from a slope, 80 m a.s.l.

Subass./variant/community	J24													J25						
	J24.1			J24.2.a		J24.2.b			J24.2.c		J24.2.d			J25.1			25.2	J25.3		
No. of relevés			4		8			12		6		6	36			6			4	11
F% (F) or average F% (av.)	F	av.		av.		F	av.		av.		av.			F	av.		F	av.		
Analysis no.	33					31								45			194			
Ditrichum flexicaule	50	30	4	61	8	80	78	12	73	6	68	6	36	10	23	3	30	73	4	8
Distichium capillaceum	100	53	4	64	8	90	78	11	58	6	55	6	35		33	3	10	85	4	8
Tortella fragilis		3	1	36	7	10	20	10	12	5	13	4	27	20	22	4		18	3	7
Hypnum bambergeri				43	6		43	11	80	6	28	3	26		2	1		53	4	5
Bryoerythrophyllum recurvirostre	10	5	2	26	5		9	5	5	1	13	3	16		15	2		20	2	4
Cladonia pyxidata		3	1	11	3		18	5	3	2	2	1	12		28	2	30	10	1	4
Orthothecium intricatum				38	5		13	4	32	3	10	1	13					15	2	2
Campylium stellatum	30	8	1	14	3		5	3	10	1	15	2	10		2	1		8	1	2
Encalypta procera		3	1	8	4		8	4	3	1			10					10	2	2
Pohlia cruda	60	38	3	1	1				2	1			5	90	52	5	10			6
Encalypta longicollis				5	1		11	6	10	2	5	1	10		2	1				1
Isopterygium pulchellum		5	1	8	2		3	1			2	1	5	20	33	3				3
Cyrtomnium hymenophylloides				4	1		16	5	3	2			8		2	1		13	2	3
Myurella julacea				5	2		3	3					5		22	2	20	5	1	4
Encalypta rhabdocarpa				8	3		3	2			3	2	7					10	3	3
Scorpidium turgescens				18	3		5	2			12	2	7					10	1	1
Tomenthypnum nitens	10	8	2	6	2	10	1	1	5	1	13	1	7					3	1	1
Amphidium lapponicum							2	1	8	2			3		20	3		3	1	4
Oncophorus wahlenbergii						20	12	4					4	50	20	2				2
Blepharostoma trichophyllum				1	1		3	2					3	60	17	3				3
Sanionia uncinatus		10	2	1	1		1	1					4		2	1				1
Solorina octospora				3	2		1	1	7	1			4		2	1				1
Nostoc sp.		3	1				1	1					2					25	2	2
Calliergon trifarium				1	1						20	2	3		3	1				1
Polytrichastrum alpinum	20	5	1										1	30	12	2				2
Cetraria nivalis				3	1		1	1					2				10			1
Thamnolia vermicularis				1	1								1				10	3	1	2
Platydictya jungermannioides		3	1										1		2	1				1
Hypnum revolutum	60	40	3	14	3		8	4	5	1	3	1	12							
Tortula ruralis	20	20	2	3	1	10	3	4			3	1	8							
Drepanocladus aduncus				6	1				8	2	28	3	6							
Brachythecium groenlandicum		3	1	5	2		7	1			15	2	6							
Mnium thomsonii		3	1	3	1		1	1	3	1			4							
Schistidium apocarpum							5	3	3	1			4							
Meesia uliginosa							2	2	5	2			4							
Orthothecium chryseum							3	3			2	1	4							
Didymodon asperifolius				13	2						2	1	3							
Brachythecium turgidum				1	1						13	2	3							
Drepanocladus intermedius											15	2	2							
Physconia muscigena				1	1		1	1					2							
Aulacomnium turgidum														100	48	4	30			5
Polytrichum strictum														70	23	2	90			3
Polytrichum juniperinum															25	2				2
Dicranum spadiceum														20	10	2				2
Anastrophyllum minutum														40	8	2				2
No. of species, range	6-12			9-16		7-17			7-12		8-11			3-19			15	10-11		
No. of species, average	9			12		10			10		9			12				11		
Summa F%, range	120-430			315-660		220-690			390-470		330-550			140-1030			410	330-510		
Summa F%, average	290			473		428			415		402			510				418		

Table 19. Ass. Rhododendro-Vaccinietum microphyllae, cryptogams.

In one relevé of the comunity only: *Arnellia fennica* (the only occurrence in the 73°N area), *Aneura pinguis, Ceratodon purpureus, Cirriphyllum cirrosum, Encalypta brevicollis, Fissidens osmundoides, Odontoschisma macounii, Orthothecium intricatum,* and, in the typical relevé of subtype of *Carex scirpoidea: Minuartia biflora* (5%), and in that of subtype of *Empetrum hermaphroditum: Poa alpina* (5%) and *Peltigera rufescens* (20%). The cryptogam analyses are given in Table 19.

Tofieldia coccinea-Carex bigelowii community (J25)

Subtype of *Carex rupestris* (J25.1)

14 grass-heath relevés from the inland and middle fjord area at 73°N were divided into three groups, each characterizing a certain area with specific pedologic and climatic conditions. Differential species are *Carex bigelowii, Tofieldia coccinea,* and *Luzula confusa*, preferential species are *Rhododendron lapponicum* and *Carex capillaris.* Average sum of F% of graminoids range be-

tween 42 and 47 in the three groups, of dwarfshrubs between 18 and 32.

Six relevés from Kap Hedlund are from almost level ground, 50-450 m a.s.l. The soil is acid: pH of the topsoil 4.8-5.9 (6 anal.), subsoil 4.7-6.0 (5 anal.), reflected in the differential or preferential species against the two following subtypes: *Hierochloë alpina, Luzula confusa, Carex bigelowii*, and *Aulacomnium turgidum*. A typical relevé (Tables 18-19, anal. 45) is from level ground 250 m a.s.l., pH of topsoil 5.0, subsoil 5.5. In one relevé only: *Campanula uniflora, Carex norvegica*, and in only one circle in one relevé: *Poa arctica, Anthelia juratzkana, Dicranum elongatum*, and *Stereocaulon paschale*. A dry *Dryas-Carex supina* ssp. *spaniocarpa* heath on dolerite outcrop on the west side of Traill Ø is a representative of the subtype (Elkington 1965, anal. 15).

Subtype of *Vaccinium microphyllum* (J25.2)
Four relevés from S and E facing slopes on Traill Ø are intermediate between the subtype of *Carex rupestris* (J.25.1) and the subtype of *Euphrasia frigida* (J25.3). The ground in three of the relevés seems fairly dry. pH of topsoil 5.5 and 6.5 (2 anal.). The fourth relevé from a S-slope with oozing water in tiny furrows is more humid, reflected in the occurrence of *Juncus biglumis, J. castaneus, J. triglumis*, and *Kobresia simpliciuscula*. Apart from the absence of *Kobresia simpliciuscula* it is rather a representative of the subtype of *Euphrasia frigida*. In one relevé only: *Draba fladnizensis* and *Juncus castaneus*. Cryptogam analysis is only available from one of the drier relevés (Table 19 anal. 194). Not included in the table: *Dicranum fuscescens* (60%), *Stereocaulon alpinum* (10%), and the only occurrence in the 73°N area of *Oreas martiana* (60%) and *Abietinella abietina* (20%).

Subtype of *Euphrasia frigida* (J25.3)
Four relevés from a ridge, 200 m a.s.l. on Ella Ø are slightly humid throughout the

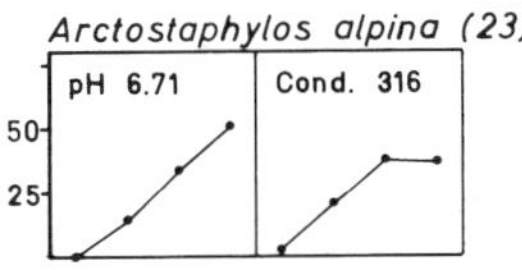

Fig. 20. pH and conductivity for *Arctostaphylos alpina*.

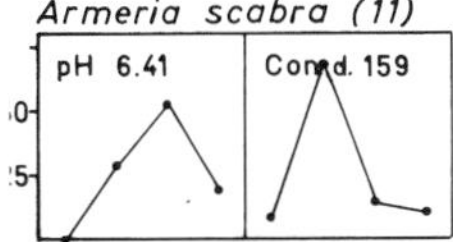

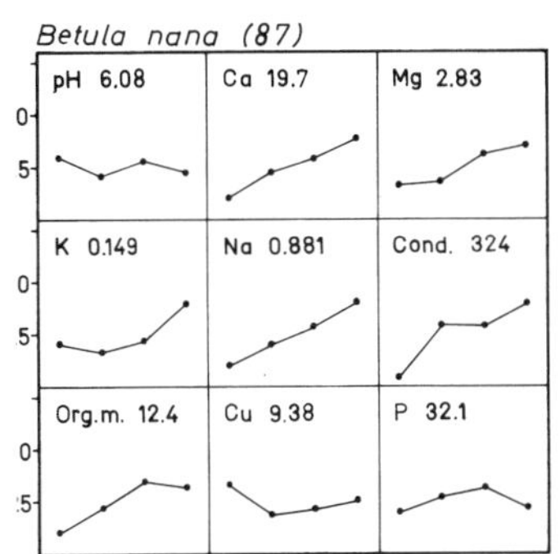

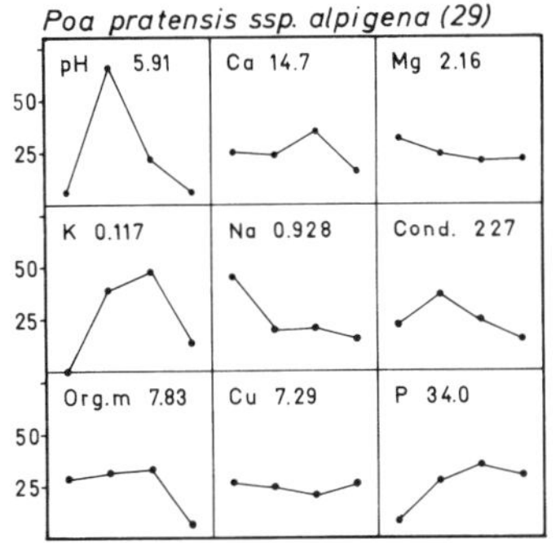

Fig. 21. Soil characteristica for three species common in ass. Rhododendro-Vaccinietum microphylli.

summer, as indicated by e.g. *Kobresia simpliciuscula, Pedicularis flammea, Juncus biglumis, J. triglumis. Armeria scabra* attains its highest occurrence in the 73°N area in this community. pH of the topsoil 6.7-7.7 (4 anal.).

Further species occurring in one relevé of the community only: *Calamagrostis purpurascens, Poa glauca, Bartramia ityphylla, Campylopus schimperi, Cetrariella delicei, Drepanocladus brevifolius, Hymenostylium recurvirostrum, Hypnum callichroum, Polytrichum piliferum*.

6. All. Dryadion integrifoliae Ohba ex Daniëls 1982

The middle-arctic dwarfshrub heaths and fell-fields of this alliance in Northeast Greenland have been divided into
1) Dwarfshrub heaths, subdivided into
Cassiope heaths on acid soil
Mixed dwarfshrub heaths on mesic, weakly acid soil
Betula nana-Dryas octopetala vegetation on dry, sunny calcareous slopes
2) Dwarfshrub and graminoid heaths, subdivided into
Open grass-heaths on dry, neutral-basic soil
Dwarfshrub and graminoid heaths on dry, weakly acid soil
Dry, graminoid *Dryas* heaths and fell-fields

6.1. Middle-arctic dwarfshrub heaths

6.1.1. Cassiope heaths on acid soil

According to Sørensen (s.a.) groups G and H comprise the "meagre, species poor, humus forming *Cassiope* heaths. pH is lower than in any other plant community". They are almost exclusively met with on gneissic bedrock at the head of the fjords in the 73°N area but are missing on the neutral-basic sediments and basalt in the middle fjord area, and only fragmentarily developed at the outer coast. The most characteristic associate species are *Huperzia selago*, almost only found with *Cassiope*, and *Poa arctica* and *Luzula confusa*. Further north, *Huperzia* is exclusively associated with *Cassiope* heaths, most frequently on crystalline rocks (Gelting 1934, Bay 1992). In the inland *Pyrola grandiflora* is almost constant, and *Betula nana* very common (group G); both species are absent at the coast, where *Stellaria longipes* s.l. is constant (group H).

Sørensen "considers this poor vegetation type as a high-arctic substitute for the Alliance Loiseleurieto-Arctostaphylion Nordhagen (1943, p. 59)". Daniëls (1982) in discussing the *Cassiope* heaths in the southern arctic regions, mainly in Greenland, points out that for floristical and ecological reasons communities dominated by *Cassiope* should be assigned to different syntaxa. According to him, the mainly chionophytic communities on acid soil should be assigned to Phyllodoco-Myrtillion Nordh. 1943, and based on analyses in Böcher (1933) from the outer coast at Kap Evert (69° 22'N) on the Blosseville Coast the association Cassiopetum tetragonae Böcher 1933 em. Daniëls 1982 is proposed. At the southern limit of *Cassiope* in East Greenland, just north of Ammassalik, the species is only found inland, almost exclusively at high altitudes on N and E facing slopes. From this area Daniëls (1982) gives seven relevés from acid grounds (pH 4-5½) on steep, mainly northern slopes 260-720 m a.s.l. as examples of the subass. typicum. *Harrimanella hypnoides* is here a differential species within the alliance, *Salix herbacea* a constant companion. Neither of these species occur in the *Cassiope* heaths in the area at 73°N and further north. Elvebakk (1985) considers *Cassiope tetragona* communities as differential syntaxon between his Southern and Middle Arctic Tundra Zone, being the "zonal vegetation" of the latter.

The middle-arctic *Cassiope* heaths of East Greenland differ from Cassiopetum tetragonae, and according to Daniëls (pers. comm. 1996) they probably represent a new association provisionally described here as Saliceto-Cassiopetum tetragonae, a North American, arctic, chionophytic association of Dryadion integrifoliae. It is characterized by the regional faithful *Cassiope tetragona* and *Huperzia selago*. It is vicarious to the European arctic-alpine Dryado-Cassiopetum, differentiated by the regional species like i.a. *Salix arctica, Dryas integrifolia*, and *Pedicularis hirsuta*.

Subassociation	G21																
Subass./variant/community	G21.1		G21.2		G21.3		J30.1	H22		Z12					Z13		
Area	H		H		H		T	HH		Z				CF	Z		
Elevation, m	50-600		20-60		40-150		275	0-100		35-45				130	25-110		
No. of relevés		*6*		*5*	*4*	*15*			*3*			*6*					*5*
F% (F) or average F% (av.)	F	av.	F	av.	av.		F	av.		F	av.		F	F	F	av.	
Analysis no.	39		62				227			6			4	2	118		
Cassiope tetragona	100	87	100	85	100	*15*	60	100	*3*	100	98	*6*	80	100	100	98	*5*
Salix arctica	95	68	65	61	76	*15*	35	63	*3*	100	93	*6*	100	100	100	96	*5*
Vaccinium microphyllum	100	98	100	96	5	*14*	100	13	*1*	80	18	*5*		20	100	100	*5*
Polygonum viviparum	90	57	55	22	8	*12*	65	37	*2*	40	32	*6*	40	67	30	12	*3*
Luzula confusa	50	28	40	30	49	*14*	10	13	*1*	20	23	*6*	40	40		10	*1*
Poa arctica	50	31	40	23	13	*11*	10	5	*1*	10	28	*5*	50	40	10	16	*3*
Dryas octopetala/sp.	70	33		1	3	*6*	50	52	*2*	30	35	*6*	20	93	10	8	*4*
Luzula arctica	10	8	5	3	13	*7*		43	*3*	30	33	*6*	30	53	10	2	*1*
Carex rupestris	100	48		1		*5*	95	57	*2*		8	*4*	10	93	20	4	*1*
Carex bigelowii	100	70		22	8	*10*	40							86		32	*3*
Hierochloë alpina	5	9	25	6	9	*8*				20	32	*6*		20	10	6	*2*
Cardamine bellidifolia	25	9	15	16	24	*8*					7	*3*	30	7		4	*1*
Oxyria digyna	45	10		1	13	*4*					+	*1*		47	10	2	*1*
Stellaria longipes s.l.								35	*3*	10	17	*6*	70	60	20	10	*2*
Saxifraga oppositifolia	65	20			6	*6*	35	18	*2*		2	*1*		7			
Silene acaulis	10	4	5	2	1	*6*	15	3	*1*		2	*1*		40			
Carex misandra		2				*1*		5	*2*	10	5	*5*		27			
Huperzia selago				11	8	*3*		2	*1*	10	5	*2*					
Pyrola grandiflora	5	35	80	42	50	*14*											
Pedicularis hirsuta		6		2	29	*9*					5	*3*	20				
Kobresia myosuroides		3	5	1		*3*		2	*1*					7			
Festuca brachyphylla/s.l.	5	6			1	*4*					5	*3*	60				
Saxifraga cernua		1				*1*		3	*1*				10	7			
Equisetum arvense				4		*1*					3	*1*					
Papaver radicatum					3	*1*					5	*4*	20	47			
Draba lactea					5	*1*	5				2	*1*	10				
Betula nana	80	52	80	60	1	*10*	100										
Tofieldia coccinea		1		13	18	*6*										6	*2*
Pedicularis lapponica		14	35	25		*7*	50										
Eriophorum triste		1	2			*3*									50	46	*3*
Draba glabella		6	5	1		*3*											
Carex capillaris		13				*4*											
Empetrum hermaphroditum			100	77		*5*									100	98	*5*
Chamaenerion latifolium		4		13	1	*4*	5	2	*1*								
Arctagrostis latifolia										30	23	*5*	90			20	*2*
Alopecurus alpinus										40	17	*5*		40		6	*1*
No. of species, range	13-18		12-19		9-18		16	8-14		12-21			16	21	7-13		
No. of species, average	17		15		13			11		17					10		
Summa F%, range	575-1005		450-835		280-640		680	410-490		400-650			680	1001	540-640		
Summa F%, average	728		626		439			460		498					556		
Aulacomnium turgidum	80	70	60	54	45	*15*											
Dicranum spadiceum	90	67	50	58	13	*11*											
Sanionia uncinatus	20	25	70	40	35	*11*											
Oncophorus wahlenbergii		32		22	43	*11*											
Pohlia cruda	40	22	50	20	33	*11*											
Hypnum revolutum		25		2	45	*7*	10										
Pohlia nutans		18		6	43	*6*											
Blepharostoma trichophyllum		15	10	16	25	*6*											
Polytrichum piliferum		3		4	20	*4*											
Brachythecium groenlandicum		3				*2*	20										
Hypnum callichroum	70	12				*1*											
Polytrichum strictum	10	43	50	22		*7*											
Anastrophyllum minutum		8	10	14		*4*											
Isopterygiopsis pulchella		12	20	4		*4*											
Distichium capillaceum		17	10	2		*3*	50										
Ceratodon purpureus		2	10	4		*3*											
Ditrichum flexicaule	10	7			20	*5*	40										
Polytrichastrum alpinum		8			28	*4*											
Cladonia pyxidata		5			8	*2*	10										
Tomenthypnum nitens		2			3	*2*	10										
Tortella fragilis		2				*1*	60										
Bryoerythrophyllum recurvirostre			10	2		*2*											
Timmia austriaca				2	25	*4*											
Dicranum fuscescens				8	18	*2*											
Dicranum scoparium					23	*1*											
No. of species, range	9-17		5-14	8-12		12											
No. of species, average	12		10	11													
Summa F%, range	320-680		160-550	410-580		310											
Summa F%, average	463		370	520													

Table 20. Ass. Saliceto-Cassiopetum tetragonae subass. pyroletosum grandiflorae (G21) and intermediate vegetation types (Z12-13).

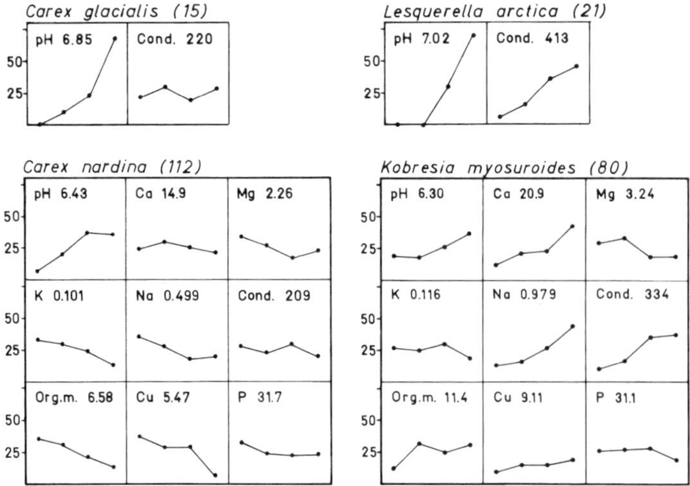

Fig. 22. Soil characteristica for four species common in ass. Carici-Dryadetum integrifoliae subass. caricetosum rupestris.

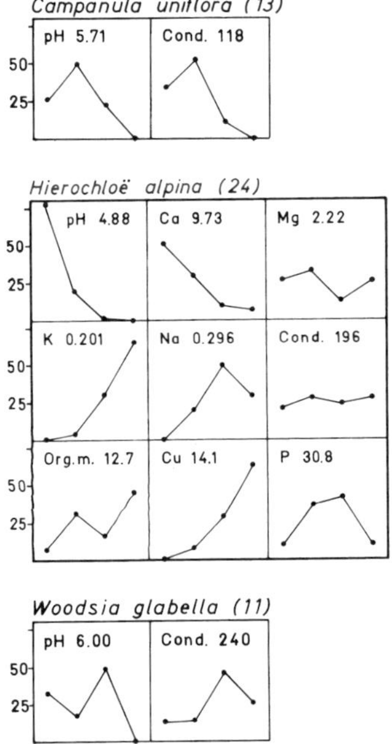

Fig. 23. Soil characteristica for three species common in ass. Carici-Dryadetum integrifoliae var. of *Carex capillaris.*

Ass. Saliceto-Cassiopetum tetragonae Daniëls & Fredskild ass. nov. prov. (G21, H22, J30, Z12-13)

Subass. pyroletosum grandiflorae subass. nov. prov. (G21)

Typical variant var. nov. prov. (G21.1)

Five lowland relevés (max. 300 m a.s.l.) from Kap Hedlund are dominated by *Vaccinium uliginosum* ssp. *microphyllum, Cassiope tetragona,* and *Betula nana.* The latter is missing in a relevé 600 m a.s.l., in which, however, *Dryas octopetala* is co-dominant. Differential species against Cassiopetum tetragonae subass. typicum are: *Betula nana, Carex capillaris, C. rupestris, Dryas* sp., *Hierochloë alpina, Pyrola grandiflora,* and *Saxifraga oppositifolia.* The most frequent species of the usually thick moss carpet are the differential species *Aulacomnium turgidum* and *Dicranum spadiceum.* pH is 4.5-5.1 (5 anal.), reflected in the ecological preferences of some of the common species (Figs 12, 17-20, 23). The type relevé (Table 20, anal. 39) is from a N-facing slope, 125 m a.s.l. In one relevé only: *Cerastium arcticum, Equisetum variegatum, Cetraria nivalis, Peltigera rufescens,* and further *Dicranum acutifolium,* which has its only occurrence in the 73°area here.

The phytosociological units of pyroletosum grandiflorae have affinity to the neutro-basophilous *Betula nana-Cassiope tetragona-Vaccinium uliginosum* heaths (J24), exclusively occurring on calcareous soils on Ella Ø, and the neutrophilous *Cassiope tetragona-Vaccinium uliginosum* heaths on Traill Ø (J30), both in the 73°N area.

Variant of *Empetrum hermaphroditum* var. nov. (G21.2)
Five lowland relevés from slightly sloping ground on Kap Hedlund differ in the co-dominating *Empetrum hermaphroditum*, and in the far less frequent *Carex rupestris* and *Dryas octopetala*. A typical relevé (anal. 62) is from a SE-facing slope, 20 m a.s.l. pH is 4.3-5.0 (5 anal.). In one relevé only: *Melandrium affine, Tofieldia palustris, Dicranum elongatum*, and further *Desmatodon latifolius* and *Tritomaria quinquedentata*, both having their only 73°N area occurrence here.

Variant of *Pedicularis hirsuta* var. nov. (G21.3)
Four lowland relevés from N-facing slopes on Kap Hedlund differ from the two preceeding units in the very few dwarfshrubs other than *Salix arctica* and the all dominating *Cassiope tetragona. Tofieldia coccinea* is growing in three relevés. pH is 4.3-5.7 (3 anal.). In one relevé only: *Rhododendron lapponicum, Amphidium lapponicum, Hypnum bambergeri, Ptilidium ciliare.* The only occurrence of *Dicranum scoparium* in the 73°N area is in a relevé of this community.

Fragmentary communities (J30.1, H22)
A relevé of a slightly hummocky "*Betula-Vaccinium* vegetation on an E-facing slope" 275 m a.s.l. on Traill Ø was by Sørensen placed as subgroup J30.1 consisting of this only relevé, together with 27 relevés of Saliceto-Cassiopetum tetragonae of all. Dryadion integrifoliae (J30.2-5). *Betula nana* is dominant, but all missing in the 27 relevés, which rather makes it a representative of Saliceto-Cassiopetum tetragonae. It is included in Table 20, anal. 227. The main difference is in the moss layer, where i.a. *Aulacomnium turgidum* is missing. Not included here: *Campanula uniflora* (5%), *Dicranum elongatum, Polytrichum juniperinum*, and *Tortella fragilis* (60%), *Tortula ruralis* (20%) and *Encalypta rhabdocarpa, Fissidens osmundoides, Meesia uliginosa*, and *Peltigera rufescens* (10%). In the 73°N area *Dicranum elongatum* was only seen here and in the var. of *Empetrum hermaphroditum* (G21.2).

Towards the coast the *Cassiope* heaths are few, small, and depauperate. Three lowland relevés from the Hold with Hope area may illustrate this (Table 20, H22). *Pyrola grandiflora, Carex bigelowii*, and *Betula nana* are all absent, *Luzula arctica* is more frequent than in the inland units, and *Stellaria longipes* s.l. is new.

Transitional heaths (Z12-13)
Two heaths types in the Zackenberg area are intermediate between subass. pyroletosum grandiflorae (G21) of Saliceto-Cassiopetum tetragonae and subass. betuletosum nanae (F15) of Arctagrostio-Eriophoretum tristis.

Intermediate type of *Arctagrostis latifolia* (Z12)
Some of the dwarfshrub heaths in the Zackenberg lowland are all dominated by *Cassiope* (Fredskild & Bay 1993, Table 1, anal. 1, 2, 5, 6, 7, and 12). Most typically, such heaths occur here, as very often in Northeast Greenland, as a belt on S-facing slopes between an earlier snowfree zone above, and a later melting snowdrift below. A common zonation from above is: abrassion flat, open *Dryas-Carex nardina* vegetation, open *Vaccinium uliginosum* heath with some *Dryas* and *Cassiope, Cassiope* heath, open *Salix arctica* snowbed vegetation, and, finally, a snowbed vegetation without *Salix arctica*. This type can be found inland to c. 78°N, yet with a decreasing number of *Vaccinium uliginosum*. Differential species against pyroletosum grandiflorae are *Arctagrostis latifolia, Alopecurus alpinus*, and *Stellaria longipes* s.l. On the contrary, *Betula nana* (N-limit 76°) and *Pyrola grandiflora* (N-limit 74½°), characteristic of pyroletosum grandiflorae in the inland of the 73°N area, do not grow in the present type. In one relevé only: *Eriophorum triste.*

The ground is almost totally covered by a moss-lichen carpet. A typical relevé

Subassociation/variant	J30.2-3						J30.4			30.5		J30.4		
Area	T	T			Z		T			T		Z	Z	Z
Elevation, m	30-550	30-500			10-150		30-600			50-600		83	102	33
No. of relevés	*6*		*5*	*11*		*5*			*10*	*6*	*16*			
F% (F) or average F% (av.)	av.	F	av.		av.		F	F	av.	av.		F	F	F
Analysis no.		159					162	163				16	9	19
Cassiope tetragona	56	100	100	*9*	50	*5*	100	100	100	98	*16*	70	20	40
Salix arctica	35	85	49	*11*	60	*5*	85	85	95	78	*16*	80	100	100
Dryas octopetala/sp.	55	85	80	*11*	86	*5*	5	20	18	68	*14*	80	70	80
Polygonum viviparum	46	100	88	*11*	50	*4*		90	37	50	*13*	50	100	10
Silene acaulis	3	35	38	*7*	2	*3*		10	11	37	*14*	20		60
Carex bigelowii	32	100	55	*9*	26	*3*			18	55	*10*		90	
Carex rupestris	43		76	*9*	46	*4*			3	39	*8*	80	10	
Luzula confusa	15		1	*1*	34	*5*	95	65	54	18	*14*		50	40
Luzula arctica	2	10	14	*4*	18	*3*	60	35	38	8	*12*		20	
Stellaria longipes s.l.	3			*1*	20	*4*	30	5	16	3	*7*	10	30	30
Pedicularis hirsuta		20	5	*2*	4	*3*		10	4	8	*7*	20	+	20
Poa arctica	3			*1*	24	*4*	95	10	22		*6*	10	60	20
Festuca brachyphylla/s.l.	3			*2*	4	*3*			1		*1*	40	30	10
Carex glacialis			19	*2*										
Vaccinium microphyllum	99	100	83	*11*	78	*5*								
Saxifraga oppositifolia	6	100	68	*6*	2	*1*	60	40	38	77	*14*			
Carex misandra	2	10	33	*7*	4	*2*		60	42	59	*12*			
Carex nardina	9		31	*7*	8	*2*		50	9	9	*8*			
Papaver radicatum	2		1	*3*	4	*3*	5	5	8		*6*			
Arctagrostis latifolia			5	*1*	22	*3*			4	5	*5*			
Cardamine bellidifolia			1	*1*	+	*1*	25	10	7		*7*			
Carex capillaris	6		1	*3*	+	*2*			1	3	*2*			
Juncus biglumis		5	3	*2*	2	*1*			1	11	*2*			
Equisetum variegatum			2	*1*	4	*1*			1		*1*			
Kobresia myosuroides	8		27	*4*	2	*1*						10		
Huperzia selago	2		4	*3*				15	14		*7*			
Carex scirpoidea		15	6	*3*						3	*2*			
Poa pratensis ssp. *alpigena*		15	21	*3*						4	*1*			
Arenaria pseudofrigida			7	*2*						1	*1*			
Saxifraga cernua	2	10	2	*2*			10	10	10	4	*9*	10		
Oxyria digyna		5	1	*1*			60	10	13	12	*9*		10	
Equisetum arvense					18	*2*								
Poa glauca					4	*2*								
Draba lactea					2	*2*	5		3	3	*4*			+
Hierochloë alpina					6	*4*							30	30
Alopecurus alpinus					2	*2*							30	
Cerastium arcticum							10		3	1	*3*			
Draba glabella									1		*1*			+
Potentilla hyparctica													40	40
No. of species, range	10-12	16	13-22		14-22		16	19	9-20	7-21		13	16	17
No. of species, average	11		16		17				16	14				
Summa F%, range	385-510	795	540-1180		360-790		670	640	320-890	315-1005		480	690	520
Summa F%, average	432		821		584				571	658				

Table 21. Ass. Saliceto-Cassiopetum tetragonae, phanerogams. Subass. typicum (J30.2-3) and subass. luzuletosum (J30.4-5).

(Table 20, anal. 6) is from almost level ground, 37 m a.s.l. The dense moss layer is partly covered by large amounts of *Cassiope* litter. The most frequent lichens are *Cetrariella delisei, Cetraria cucullata,* and *Stereocaulon alpinum.* Further to these, an analysis from the same site (Hansen, Table 1, anal. 5, in Fredskild & al. 1995) includes *Ochrolechia frigida, Cladonia mitis, C. amaurocraea,* and *C. stricta.* The site of another relevé with a similar vegetation analyzed by Hansen (l.c. anal. 4) is dominated by *Ochrolechia frigida, Cetraria cucullata,* and *Stereocaulon alpinum.* Further species are *Cetrariella delisei, Cladonia borealis, Cetraria islandica, Psoroma hypnorum, Candelariella placodizans, Cetraria muricata,* and *Cladonia amaurocraea.*

A relevé (anal. 4) from level ground, 39

m a.s.l. is characterized by more frequent *Arctagrostis latifolia*, *Stellaria longipes*, and *Festuca brachyphylla*. Micropolygons either consist of fresh soil or they are covered by organic crust. Most of the ground is covered by lichens, including *Stereocaulon* sp. and *Peltigera leucophlebia*, and mosses.

In a relevé from a 5-10° SE-facing slope on moist silt, 130 m a.s.l., at Clausen Fjord, 77°32' N the degree of cover for *Cassiope* is 45%, *Vaccinium uliginosum* 5%, *Dryas* 4%, *Salix arctica* 1%, all other species less (Table 20, anal. 2). Dead plant material cover 30%. In this area a *Cassiope* heath may cover several hundred m^2, whereas further north it is usually only a few m^2 wide.

Intermediate type of *Empetrum hermaphroditum* (Z13)

In the Zackenberg valley dense *Empetrum hermaphroditum-Cassiope-Salix arctica-Vaccinium microphyllum* heaths mainly occur west of the river on level or S-SW-facing slopes. The ground is mostly almost covered by mosses, lichens, and litter, yet in one of the relevés (Fredskild 1996, anal. 114) the moss cover was very open and no lichens seen because of a very dense cover of dwarfshrubs. Five lowland relevés (l.c. anal. 114-116, 118, 130) are representatives of the type, which differs from the closely related var. of *Empetrum hermaphroditum* (G21.2) in the absence of *Betula nana*, *Pyrola grandiflora*, and *Pedicularis lapponica*. Northwards it is found to the inland at 75°18'N (Bay & Fredskild 1990).

A typical relevé (Table 20, anal 118) is from a slightly S-facing slope, 40 m a.s.l. with the following succession from above: a *Dryas* heath, a *Vaccinium-Cassiope* heath, a 1-1½ m wide belt with the relevé, and on the almost level ground a mossy *Salix arctica-Vaccinium microphyllum* heath. *Peltigera aphthosa* is the all dominating lichen. In one relevé only: *Draba arctica* (10%), *Rhododendron lapponicum* (+). Other lichens seen in the relevés: *Cetraria cucullata*, *C. nivalis*, *Psoroma hypnorum*, *Stereocaulon alpinum*.

Subassociation/variant	30.2	30.3	30.4	30.5	
No. of relevés	*3*	*4*	*7*	*3*	*17*
Average F% (av.)	av.	av.	av.	av.	
Ditrichum flexicaule	10	65	57	70	*17*
Cladonia pyxidata	17	15	31	27	*13*
Distichium capillaceum	13	40	27	37	*12*
Cetraria nivalis	40	25	9	17	*8*
Fissidens osmundoides	10	13	6	10	*5*
Peltigera rufescens	23				*2*
Polytrichum strictum	30	20			*4*
Sanionia uncinatus	13		16		*8*
Aulacomnium turgidum	30		19		*3*
Tortula ruralis	3		1		*2*
Anastrophyllum minutum	3			3	*2*
Pohlia cruda	17		19	13	*10*
Isopterygiopsis pulchella	17		11	17	*8*
Dicranum spadiceum	30		40	23	*7*
Polytrichum juniperinum	27		17	33	*7*
Tortella fragilis		23	16	40	*10*
Cetrariella delicei		20	59	57	*9*
Cesia concinnata		3	26	13	*7*
Tomenthypnum nitens		28	3	7	*4*
Odontoschisma macounii		3	10		*2*
Hypnum revolutum		5	3		*2*
Hypnum bambergeri		45		27	*3*
Oncophorus wahlenbergii		3		7	*2*
Meesia uliginosa		3		7	*2*
Bryoerythrophyllum recurvirostre		3		2	*2*
Stereocaulon alpinum			19	30	*6*
Polytrichastrum alpinum			19		*4*
Amphidium lapponicum			4	13	*3*
Schistidium apocarpum			1	3	*2*
Racomitrium canescens			1	3	*2*
No. of species, range	10-12	7-13	9-25	12-15	
No. of species, average	11	11	14	13	
Summa F%, range	360-420	300-510	230-640	380-610	
Summa F%, average	380	420	446	500	

Table 22. Ass. Saliceto-Cassiopetum tetragonae, cryptogams. Subass. typicum (J30.2-3) and luzuletosum (J30.4-5).

6.1.2. Mixed dwarfshrub heaths on mesic, weakly acid soil

Cassiope tetragona dominates 25 of 27 relevés of two subassociations from Traill Ø. Differential species against pyroletosum grandiflorae (G21.1) are *Carex misandra*, *C. rupestris*, and *Dryas octopetala*. *Saxifraga oppositifolia* and *Silene acaulis* are more frequent, *Poa arctica* less frequent, and *Betula nana*, *Hierochloë alpina*, and *Pyrola grandiflora* are missing. The soil is less acid. Phanerogams and cryptogams are presented in tables 21-22. *Gymnomitrion concinnata* is frequent, occurring in 7 of 17 relevés. Otherwise, it was only met with in two relevés of Phippsietum algidae-concinnae subass. oxyrietosum digynae (B4).

Area	EY												
Elevation, m	100	250	140	300	500	80	250	140	200	250	200	av.	
Analysis no.	84	303	296	286	330	34	110	301	120	304	118	F%	*no.*
Dryas octopetala	100	100	55	75	100	85	65	55	80	100	100	87	*11*
Betula nana	30	100	100	95		100	100	100	100	100	95	84	*10*
Carex rupestris	100		100	100	100	75	75	25	85		100	69	*9*
Saxifraga oppositifolia	95	10	10	95		40		10	5	15	85	28	*9*
Carex nardina	85		35	35	10	5		10	30	15	30	23	*9*
Silene acaulis	20	10	5	15	5	5	5					6	*7*
Pyrola grandiflora		40	100		95	100		90	100			48	*6*
Polygonum viviparum	75	65		10		15			10	20		18	*6*
Salix arctica		70	20	5		10				35	40	12	*6*
Draba glabella	15	20			15	45	15		20			12	*6*
Poa glauca	5	10				30	5	5	5			5	*6*
Arctostaphylos alpina	80	15	15	5	5							11	*5*
Minuartia rubella	25				5	20	20				5	7	*5*
Poa alpina		30				5	50			15		9	*4*
Arenaria pseudofrigida	20					10	15	15				5	*4*
Melandrium triflorum/affine		5				15				15	5	4	*4*
Saxifraga cernua	20						25					4	*2*
Draba nivalis	10							25				3	*2*
Minuartia biflora		10								5		1	*2*
Lesquerella arctica	45											4	*1*
Poa pratensis ssp. *alpigena*		35										3	*1*
Cerastium arcticum						35	15	10	20			7	*4*
Saxifraga nivalis						20	40		10	5		7	*4*
Festuca rubra								100				9	*1*
Draba arctica/cinerea							75					7	*1*
Saxifraga caespitosa						30						3	*1*
Number of species	18	15	12	9	9	19	13	12	11	10	8		
Summa F%	750	530	460	435	340	655	505	450	465	325	460		
Ditrichum flexicaule	100		70	10		60			30		60	41	*6*
Encalypta rhabdocarpa	40		10	40	50	40	20					25	*6*
Distichium capillaceum	40			90		85			30		20	24	*5*
Tortella fragilis	70			40	17	60					10	23	*5*
Tortula ruralis			90		17	5	70		20		20	23	*5*
Cladonia pyxidata				80		50	10		20		20	23	*5*
Bryoerythryphyllum recurvirostre	80					15					10	13	*5*
Stegonia latifolia					17						10	3	*2*
Myurella julacea	20											3	*1*
Hypnum revolutum			80									10	*1*
Ceratodon purpureus			20									3	*1*
Amphidium lapponicum				20								3	*1*
Peltigera rufescens							70		20			11	*2*
Polytrichum alpinum						20			30			6	*2*
Physcia muscigena						5	10					2	*2*
Pohlia cruda						65						8	*1*
Number of species	7		7	10	7	12	6		6		9		
Summa F%	370		320	340	152	470	190		150		180		

Table 23. *Arctostaphylos alpina-Betula nana* community (I23).

Cetraria nivalis is character species, *Cetrariella delisei* preferential, being most frequent here, in caricetosum rupestris (J28-29), and in two units (B5.1-2) of oxyrietosum digynae.

Subass. luzuletosum subass. nov. prov. (J30.4)

Ten relevés are from snow protected but fairly early snow free zones on only slightly sloping sites. An example of a zonation related to the snow duration is from a slope, 600 m a.s.l. Latest free of snow is a species rich "snowbed" of oxyrietosum digynae (Table 5, anal. 161), followed by a "*Cassiope* snowbed" (Table 21, anal. 162, the type relevé of the subass.), which is replaced by an earlier snow free "more open, more vigorously flowering (June 20) *Cassiope* heath" (Table 21, anal. 163), more dry, as indicated e.g. by *Carex nardina*. Cryptogam analyses are available from seven of the relevés, but not from these two (Table 22). *Stereocaulon alpinum* is a characteristic species, otherwise found in only two relevés in the 73°N area. pH is 4.6-6.5 (5 anal.). In one relevé only: *Draba adamsii* (15%), *Saxifraga tenuis* (10%), (both in anal. 162), *Minuartia rubella* (10%, in anal. 163), and further *Draba glabella, Saxifraga foliolosa, Bartramia ityphylla, Conostomum tetragonum, Encalypta rhabdocarpa, Hypnum callichroum, Myurella julacea, Physconia muscigena, Polytrichum piliferum, Ptilidium ciliare, Psilopilum cavifolium, Solorina octospora*, and *Timmia austriaca*.

Three lowland relevés from 8°-16° SW-facing slopes at Zackenberg (Table 21, right) are representatives of the subassociation, only slightly deviating in the occurrence of *Potentilla hyparctica* and *Hierochloë alpina* in two, and the missing *Saxifraga oppositifolia* and *Carex misandra*. Traces of cryoturbation visible, but most of the ground covered by lichens, e.g. *Cetraria islandica* and *Candelariella vitellina*, mosses, and organic crust. In one relevé only: *Melandrium triflorum* and *Minuartia biflora* (both +), and *Trisetum spicatum* (40%), all in anal. 19, which is from a 4 m wide belt below a dense, earlier snowfree *Cassiope* heath with *Salix* and *Dryas*.

A hummocky *Cassiope-Dryas-Vaccinium* heath on a raised marine bed at Mestersvig (Elkington 1965, anal. 5) is a representative of the subass.

Variant of *Dryas octopetala* var. nov. (J30.5)

Six relevés, separated by Sørensen as subgroup 30.5, are very closely related to the subass., the main difference being more *Dryas* and *Carex bigelowii*. The soil is drier, often stony, pH 4.9-6.6 (5 anal.). In one relevé only: *Carex parallela, Minuartia biflora, Pedicularis flammea, Saxifraga aizoides, Orthothecium intricatum*.

Subass. typicum subass. nov. (J30.2-3)

Five relevés on slightly SE- and E-sloping sites on Traill Ø are early free of snow and the soil is drier than in subass. luzuletosum, as indicated by *Carex glacialis* and *Kobresia myosuroides*, both frequent in two relevés, by i.a. the markedly fewer *Luzula confusa, L. arctica*, and *Saxifraga oppositifolia*, and by the differential species *Vaccinium uliginosum* ssp. *microphyllum*. *Dryas* and *Carex rupestris* are more frequent. pH is 5.1-6.6 (4 anal.). The type relevé (anal. 159) is from a S-facing solifluction slope 500 m a.s.l., still on June 19 with a snow drift below the ridge. Just below this is a relevé of the subtype of *Minuartia stricta* (B5.2) of Phippsietum algidae-concinnae subass. oxyrietosum digynae, still with oozing water, then a relevé of the variant of *Carex bigelowii* (E13) of Saxifrago-Kobresietum simpliciusculae, followed by the type relevé of the present subassociation (Table 21 anal. 162), and finally in the driest zone a relevé of the variant of *Dryas octopetala* (J30.5). In one relevé only: *Cetraria islandica, Dicranum fuscescens, Encalypta procera, Orthothecium chryseum, Saelania glaucescens, Thamnolia vermicularis*.

Six relevés, separated by Sørensen as subgroup 30.2 are from slopes of different orientation, the soil sandy-stony with only a thin humus layer, pH 5.3-5.9 (4 anal.). The vegetation is markedly more open, but the species composition is very similar to that of the subassociation, and consequently they are considered representatives of this (Table 21, first column from left). In one relevé only: *Chamaenerion latifolium, Tofieldia coccinea, Aulacomnium palustre, Brachythecium groenlandicum, Bryoxiphium norvegicum, Cetraria cucullata, Dicranum fuscescens* var. *congestum*.

Likewise, five lowland relevés from SW-facing slopes at Zackenberg (Fredskild & Bay 1993, anal. 3, 8, 11, 17 and 18) are representatives of the subassociation and included in the table. In the beginning of August the soil is dried out. Lichens, i.a. *Cetraria cucullata, Cetrariella delisei, Cladonia pocillum, Stereocaulon alpinum*, and *Thamnolia subuliformis*, and mosses cover all the surface apart from small areas with slow cryoturbation. Here, the soil is covered by organic crust. In one relevé only: *Armeria scabra* and *Eriophorum triste*. E.S. Hansen has analyzed the site of one of the relevés (Fredskild & al. 1995, Table 1, anal. 3). Dominating lichens are *Stereocaulon alpinum*, *Ochrolechia frigida*, and *Cetraria cucullata*, less frequent *Alectoria nigricans*, *Cladonia pocillum*, *C. borealis*, *Cetraria muricata*, *Psoroma hypnorum*, *Rinodina turfacea*, and in only one circle: *Cetrariella delisei*, *Peltigera leucophlebia*, and *P. didactyla*.

6.1.3. Betula nana-Dryas octopetala vegetation on dry, sunny, calcareous slopes

Arctostaphylos alpina-Betula nana community (I23)

According to Sørensen (s.a.) "Group I represents a specific vegetation type, only found on Ella Ø-Ymer Ø, where it replaces herbslopes on sunny, desiccating slopes. Characteristic species are firstly *Betula nana*, often in larger but more scattered individuals compared to more mesic sites, and secondly *Dryas octopetala*. *Arctostaphylos alpina* is only exceptionally found outside this community, where it is best developed. As sociologically and ecologically it deviates so much from its occurrence in Scandinavia it is tempting to assume two genetically different races. This Betuleto-Arctostaphyletum alpinae Ass. is mostly related to All. Kobresieto-Dryadion, and in certain respects also to Saxifragion cotyledonis, but not to the Scandinavian Arctostaphylos alpina vegetations of All. Loiseleurieto-Arctostaphylion". According to Daniëls (1982) the correct name of Kobresio-Dryadion Nordh. 1936 is Caricion nardinae Hadac 1946, which in arctic America is replaced by Dryadion integrifoliae Ohba ex Daniëls 1982.

The group is highly related (QS = 54) to the species rich *Draba glabella-Lesquerella arctica* community (L39) of Cerastio-Festucetum brachyphyllae (All. Veronico-Poion glaucae) found on sedimentary bedrock, and also, yet with a lower QS (44-45) to some heath types (J26, J27, J28) of Dryadion integrifoliae on neutro-basophilous soil, all in the same area. The species composition of the 11 relevés is rather diverse, with several species being common in only one relevé, e.g. *Festuca rubra*, *Draba cinerea*, *Lesquerella arctica*, and *Poa alpina*. Because of this, all relevés are presented in Table 23, arranged after decreasing frequency and number of species of phanerogams. The five relevés with *Arctostaphylos alpina* are grouped together. The syntaxonomical position is uncertain, and in the synoptical table (Table 36) Daniëls has placed the community, including all 11 relevés, under Veronico-Poion glaucae.

Species occurring in two of twenty circles in one relevé only: *Chamaenerion latifolium*, *Euphrasia frigida*, *Kobresia myosuroides*, *Trisetum spicatum*, *Woodsia glabella*, and in only one circle: *Campanula gieseckiana*, *Carex capillaris*. Cryptogams in only one of ten circles in one relevé only: *Cetraria nivalis*, *Encalypta alpina*, *Platydictya jungermannioides*, *Tortula mucronifolia*. Cryptogam analyses are only available for eight relevés. In the 73°N area *Tortula ruralis* and *Encalypta rhabdocarpa* are most frequent here and in Cerastio-Festucetum brachyphyllae (Z14), and the latter also in Arabido holboellii-Caricetum supinae (L37). The average number of species and the average F% for phanerogams is 12 and 489, for cryptogams 8 and 272, respectively. pH of the topsoil is 5.9-7.9 (11 anal.), of deeper layers 6.1-6.8 (4 anal.). *Arctostaphylos alpina* seems restricted to circumneutral-basic soils with high conductivity (Fig. 20).

In W.Greenland *Arctostaphylos alpina* is lowarctic, with two small distribution areas at 65°-65½°N and 69½°-70°N (Fredskild 1996). In East Greenland it is restricted to the Middle arctic tundra zone between 68°43'N and 74°49'N, being fairly frequent only in the middle fjord and inland areas. It was only found once in the Zackenberg valley. If Sørensens suggestion of treating this vegetation as Betuleto-Arctostaphyletum alpinae is followed, anal. 296 is the type relevé. In this relevé only: *Armeria scabra*, *Papaver radicatum*, and *Schistidium apocarpum* (10%), and *Pedicularis lapponica* (5%).

6.2. Middle-arctic dwarfshrub and graminoid heaths

According to Sørensen (s.a.) group J includes "several physiognomically and also ecologically different associations forming a group, beyond doubt representing the all. Kobresieto-Dryadion (Nordhagen 1943 p. 573), maybe in a somewhat extented sence. The vegetation in the middle fjord area on the sedimentary rocks and partly also on the basalt is dominated by associations of this alliance. There is no exclusive species characterizing the all. since the ecological amplitude of important and dominating species like *Carex nardina, Dryas octopetala*, and *Carex misandra* is surprisingly wide in this country. As to similarity coefficient and species combination the group can be divided into three undergroups: o, p and q". Group o includes his syntaxonomical units J24-26, group p J27, and group q J28-31.

According to Daniëls (1982) all. Caricion nardinae Hadac 1946 (syn. Kobresio-Dryadion Nordh. 1936) is found in Scandinavia incl. Svalbard, whereas in Greenland it is replaced by Dryadion integrifoliae Ohba ex Daniëls 1982, which also includes J26-31. This alliance is present in arctic North America and in the whole of Greenland, yet more commonly distributed in continental areas and in the northern parts.

6.2.1. Open grass-heaths on dry, neutral-basic soil

On Ymer Ø and Ella Ø two phytosociological units characterize the dry, more or less wind-exposed sites. The one (J26) is termed "pure grass heath", whereas the other (J27) is "A very well-defined, species poor *Carex pedata* (= *glacialis*) Ass. with *Lesquerella*, the most distinct basophilous Ass. on fine-grained to clayey soil, highly influenced by solifluction in spring, (but) extremely drying out during summer. The vegetation is open, not covering the usually whitish grey, marly ground. Exclusively bound up with chalky sedimentary rocks in the Middle Fjord Area" (Sørensen s.a.).

Variant	J26			J27			
Area	EY			EY			T
Elevation, m	30-800			80-500			380
No. of relevés			*10*			*15*	
F% (F) or average F% (av.)	F	av.		F	av.		F
Analysis no.	315			116			221
Dryas octopetala	60	92	*10*	100	88	*15*	100
Saxifraga oppositifolia	75	83	*10*	85	63	*15*	50
Carex rupestris	65	89	*10*	80	67	*14*	55
Carex nardina	25	46	*8*	90	81	*15*	100
Lesquerella arctica	45	18	*6*	20	17	*11*	
Silene acaulis		16	*5*	10	7	*10*	20
Polygonum viviparum		65	*8*		11	*6*	35
Kobresia myosuroides	100	82	*10*		4	*3*	
Salix arctica	20	34	*8*		6	*5*	15
Arctostaphylos alpina		6	*3*		4	*3*	
Pedicularis flammea		6	*3*		1	*1*	
Betula nana		19	*2*		1	*1*	
Chamaenerion latifolium		2	*2*		2	*1*	
Draba arctica/cinerea	5	1	*2*		1	*1*	
Euphrasia frigida		1	*1*		5	*1*	
Braya linearis	25	3	*1*		1	*1*	
Draba glabella		1	*2*	5	1	*1*	
Pedicularis hirsuta		1	*1*		1	*1*	
Carex misandra		21	*4*				
Braya purpurascens	25	6	*3*				
Carex scirpoidea		11	*2*				
Woodsia glabella		9	*2*				10
Pyrola grandiflora		2	*2*				
Arenaria pseudofrigida		1	*2*				5
Kobresia simpliciuscula		1	*2*				
Saxifraga cernua		1	*2*				
Carex glacialis				40	31	*10*	65
Calamagrostis purpurascens					2	*2*	
Potentilla hookeriana/nivea					1	*2*	
No. of species, range	8-17			5-12			12
No. of species, average	12			8			
Sum of F%, range	445-800			165-740			470
Sum of F%, average	619			390			
No. of relevés			*8*			*11*	
Ditrichum flexicaule		63	*8*	40	42	*10*	
Distichium capillaceum		69	*7*	19	52	*10*	
Bryoerythrophyllum recurvirostre		35	*5*	50	46	*10*	
Tortella fragilis		46	*7*		11	*7*	
Encalypta longicollis		16	*3*	30	28	*7*	
Cladonia pyxidata		21	*5*	10	17	*4*	
Myurella julacea		6	*4*		2	*2*	
Hypnum bambergeri		30	*4*		1	*1*	
Stereocaulon paschale		6	*3*		2	*2*	
Amphidium lapponicum		4	*1*		10	*3*	
Encalypta procera		4	*2*		3	*2*	
Brachythecium groenlandicum		9	*2*		4	*1*	
Hypnum revolutum		3	*2*		4	*1*	
Solorina octospora		1	*1*		2	*2*	
Encalypta rhabdocarpa		6	*1*		4	*1*	
Didymodon asperifolius		24	*2*				
Tortula ruralis		11	*2*				
Campylium stellatum		9	*2*				
Cyrtomnium hymenophylloides		6	*2*				
Cetraria nivalis					3	*2*	
No of species, range	10-12			2-10			
No of species, average	11			7			
Summa F%, range	380-640			30-400			
Summa F%, average	450			258			

Table 24. Ass. Carici-Dryadetum integrifoliae subass. caricetosum rupestris.

Ass. Carici-Dryadetum integrifoliae Daniëls 1982
Subass. caricetosum rupestris subass. nov. (J26-29, J31, K32)
Differential species: *Carex rupestris* and *C. misandra.*

Typical variant var. nov. (J26)
Seven of the 10 relevés (Table 24) are below 200 m a.s.l., three from elevations between 450 and 800 m. They are dominated by the pedologically ubiquitous species *Dryas octopetala, Carex rupestris, Kobresia myosuroides,* all differential species, and by *Saxifraga oppositifolia* (Figs 6, 7, 22). The calciphilous species *Lesquerella arctica* (Fig. 22) and *Braya purpurascens* (Fig. 15) are characteristic. pH of the top-soil 6.1-7.5 (9 anal.). The zonation on the lee, S-facing slope of a lowland ridge or hillock is shown by relevés of (from below): Saxifrago-Kobresietum simpliciusculae (E12), two relevés of subtype of *Kobresia myosuroides* of the *Betula nana-Tofieldia pusilla* community (J24.2a), and just before the top of the present subass. Corresponding to the latter on the wind side of the ridge is the type relevé (Table 24, anal. 315), differing mainly in the occurrence of *Braya purpurascens* and *B. linearis*, both in 5 of 20 circles, in the more frequent *Lesquerella arctica*, and the less frequent *Carex rupestris* and *Dryas octopetala.* On the type relevé is noted "Poor cryptogam growth", and no samples were taken. Analyses of cryptogams are available from 8 relevés. In one relevé only: *Carex capillaris, Juncus biglumis, Melandrium affine, Minuartia stricta, Pedicularis lapponica, Blepharostoma trichophyllum, Ceratodon purpureus, Distichium inclinatum, Drepanocladus aduncus s.l., Encalypta alpina, Hylocomium splendens, Isopterygiopsis pulchella, Oncophorus wahlenbergii, Physconia muscigena, Pohlia cruda, Preissia quadrata, Schistidium apocarpum, Scorpidium turgescens, Tortula mucronifolia,* and *Trematodon brevicollis.*

The subassociation shows affinity to the var. of *Lesquerella arctica* (L38, QS = 49), and to the *Draba glabella-Lesquerella arctica* community (L39, QS = 51) both of Veronico-Poion glaucae. On Ole Rømer Land, 74°N, close to the ice cap, vast lowland areas, often whitish of salt efflorescences, are covered by steppe-like vegetations all dominated by *Kobresia myosuroides,* with *Braya purpurascens, B. humilis,* and *Lesquerella arctica* (Fredskild & al. 1992), sometimes with *Dryas* sp. and *Carex nardina* as co-dominants (Schwarzenbach 1960). They are related to the subass., but no relevés are given.

Variant inops is described below.

Variant	J28-29						
Area	T			HH			Z
Elevation, m	30-600			0-100			25
No. of relevés			*8*				
F% (F) or average F% (av.)	F	av.		F	F	F	F
Analysis no.	239			4	5	40	10
Saxifraga oppositifolia	100	62	*8*	60	90	70	50
Carex rupestris	100	51	*8*	100	80	30	100
Salix arctica	55	35	*8*	100	70	30	100
Dryas octopetala	100	84	*8*		100	45	90
Carex nardina	55	64	*8*			50	
Polygonum viviparum	20	41	*6*	20	20	100	90
Silene acaulis	10	14	*5*	60	20	5	10
Luzula confusa	10	3	*3*	90	30		30
Pedicularis hirsuta	15	6	*3*	10	10		10
Cerastium arcticum		2	*3*	20		10	
Draba subcapitata		4	*3*	40			
Kobresia myosuroides		6	*1*	100	10	45	10
Luzula arctica		3	*2*		10		
Carex capillaris		1	*1*		30		
Arenaria pseudofrigida		1	*1*			15	
Festuca brachyphylla /s.l.		1	*1*			10	50
Poa arctica		1	*1*				10
Carex misandra	30	11	*4*				
Saxifraga cernua		8	*4*				
Minuartia rubella		2	*2*				
Chamaenerion latifolium		7	*1*				
Potentilla hyparctica		3	*1*				10
Juncus triglumis		2	*1*				
Arnica angustifolia		2	*1*				
Draba nivalis				20			
Papaver radicatum				20			20
Draba lactea					10		
Campanula uniflora						15	
Poa abbreviata						5	
Hierochloë alpina							80
Juncus biglumis							30
Armeria scabra							30
No. of species, range	6-15			12	12	13	16
No. of species, average	11						
Summa F%, range	205-575			640	480	430	720
Summa F%, average	423						

Table 25. Ass. Carici-Dryadetum integrifoliae var. inops, phanerogams.

Variant of *Carex glacialis* var. nov. (J27)
The 15 relevés, equally spread between 80 and 500 m a.s.l. on Ella Ø and Ymer Ø, are

from more or less active solifluction soil, mostly on south facing slopes. Some are almost or totally without mosses, and the total cover of phanerogams as well as cryptogams is small, as indicated by the average sum of F%: 390 and 258, respectively. Vegetations of this variant can be found e.g. as a belt between a *Rhododendron-Vaccinium microphyllum* vegetation and a totally barren, windswept ridge. According to the field notes some of the sites are supposed to be snowfree in winter, and allready by the middle of May when some of the relevés were analysed, the depth of the active layer was measured to 30-45 cm. The soil is basic, pH 7.2-8.0 (14 anal.).

Carex glacialis is exclusive, otherwise only seen in four of 108 relevés of the alliance, in the two of these in only one circle, and outside the all. in only one circle. Dominating species are *Dryas octopetala, Carex nardina, C. rupestris*, and *Saxifraga oppositifolia*, and *Lesquerella arctica* is frequent. In the 73°N area *Encalypta longicollis* is most frequent here, and, besides, fairly common in the typical variant (J26) and in Saxifrago-Kobresietum simpliciusculae (E11). A typical relevé (Table 24, anal. 116) is from only slightly active solifluction soil on a south facing slope, 225 m a.s.l., pH 7.7. In one relevé only: *Orthothecium intricatum, Stegonia latifolia, Thamnolia vermicularis.*

A further relevé, mentioned below under var. inops, is included in the table (anal. 221).

6.2.2. Dwarfshrub and graminoid heaths on dry, weakly acid soil

According to Sørensen (s.a.) "undergroup q includes different neutrophilous-weakly acidophilous (the last two words later deleted by Sørensen) associations almost all from the basaltic area on Traill Ø, forming a parallel to the more basophilous associations in undergroup o. Against this the associations are characterized by abundant occurrences of *Luzula confusa* and *L. arctica*, and also *Pedicularis hirsuta*, whereas *Elyna* (= *Kobresia myosuroides*) and *Pedicularis flammea* are highly reduced. The undergroup includes meagre *Dryas-Carex nardina-C. rupestris* vegetations (28 and 29), *Cassiope-Vaccinium* heaths and *Salix* heaths (30), and pure grass slopes (31)".

Subassociation	J28-29	
Area	T	
No. of relevés	av.	*3*
Distichium capillaceum	53	*3*
Ditrichum flexicaule	43	*3*
Cladonia pyxidata	33	*3*
Cetrariella delicei	17	*3*
Polytrichum juniperinum	30	*2*
Stereocaulon paschale	30	*2*
Tortella fragilis	27	*2*
Myurella julacea	7	*2*
Pohlia cruda	13	*1*
Cetraria nivalis	13	*1*
Hypnum bambergeri	10	*1*
Tortula ruralis	7	*1*
Bryoerythrophyllum recurvirostre	7	*1*
No. of species, range	8-14	
No. of species, average	10	
Summa F%, range	170-500	
Summa F%, average	320	

Table 26. Ass. Carici-Dryadetum integrifoliae var. inops, cryptogams.

Var. inops var. nov. (J28-29)
Eight relevés from Traill Ø, the five between 400 and 600 m a.s.l., can be characterized as (fairly) dry, open *Dryas-Carex nardina-Carex rupestris* heaths on sandy to stony, patterned ground or solifluction soil, mostly slightly south or east sloping. Cryptogams, mainly lichens, are few, analyzed in only three relevés (Table 26). pH is 5.9-7.3 (6 anal.). Pedologically the dominating species are indifferent, being common in several associations. In the absence of exclusive and differential species the variant is considered var. inops. The most basophilous species *Lesquerella arctica* and *Braya purpurascens* are missing. A typical relevé (Table 25, anal. 239) is termed "*Dryas* solifluction soil heath", 40 m a.s.l. The soil is sandy-gravelly. No cryptogam analysis available.

In one or two circles (of twenty) in only one relevé: *Carex glacialis, C. scirpoidea, Equisetum arvense, Poa glauca, Woodsia glabella*, in one circle in only one relevé: *Campylium stellatum, Encalypta alpina, Isopterygiopsis pulchella.*

Variant	J31			K32		
Area	T			H		
Elevation, m	200-600			50-600		
No. of relevés			*5*			*7*
F% (F) or average F% (av.)	F	av.		F	av.	
Analysis no.	134			252		
Polygonum viviparum	100	100	*5*	100	100	*7*
Carex rupestris	100	97	*5*	95	99	*7*
Dryas octopetala	100	98	*5*	80	65	*7*
Salix arctica	55	65	*5*	60	48	*7*
Carex capillaris	75	59	*5*	95	44	*7*
Luzula confusa	10	35	*5*	65	29	*7*
Saxifraga oppositifolia	100	88	*5*	10	50	*5*
Carex bigelowii	95	86	*5*	100	47	*5*
Silene acaulis	40	28	*5*	5	8	*5*
Saxifraga cernua		13	*2*	20	40	*7*
Kobresia myosuroides		20	*1*	100	87	*7*
Carex misandra	100	64	*5*	5	1	*2*
Campanula uniflora		12	*1*	55	39	*5*
Juncus biglumis		6	*3*	15	6	*3*
Pedicularis hirsuta	45	21	*3*		2	*2*
Carex nardina	35	20	*3*		2	*1*
Papaver radicatum		14	*2*		11	*2*
Draba fladnizensis		13	*2*	5	3	*2*
Poa glauca		1	*2*		7	*2*
Carex scirpoidea		20	*2*		1	*1*
Pedicularis flammea		15	*1*	30	6	*2*
Poa pratensis ssp. *alpigena*		4	*1*		17	*2*
Chamaenerion latifolium		1	*1*		12	*2*
Cardamine bellidifolia		1	*1*	5	11	*2*
Poa arctica		8	*2*		1	*1*
Equisetum arvense		6	*2*		3	*1*
Minuartia rubella		1	*1*		4	*1*
Euphrasia frigida		2	*1*		1	*1*
Woodsia glabella	50	24	*4*			
Stellaria longipes s.l.		17	*3*			
Luzula arctica		2	*2*			
Hierochloë alpina				100	50	*5*
Draba glabella					16	*4*
Eriophorum triste				5	7	*3*
Potentilla hyparctica				20	15	*2*
Arctagrostis latifolia					13	*2*
Campanula gieseckiana					13	*2*
Cerastium arcticum					7	*2*
Rumex acetosella					6	*2*
Festuca brachyphylla s.l.					2	*2*
Saxifraga nivalis					1	*2*
No. of species, range	14	14-27		21	17-23	
No. of species, average		19			19	
Summa F%, range	910	900-1145		975	775-990	
Summa F%, average		986			859	

Table 27. Ass. Carici-Dryadetum integrifoliae var. of *Carex capillaris* (J31) and var. of *Hierochloë alpina* (K32), phanerogams.

A ninth relevé from a slightly east facing slope, 380 m a.s.l., which by Sørensen was included in group 28, differs significantly, i.a. in the occurrence of *Carex glacialis* in 13 of 20 circles. This rather makes it a representative of the variant of *Carex glacialis* (J27). Consequently, it is included in Table 24, anal. 221. Not included here are: *Luzula confusa* 10%, *Poa glauca* 5%.

Three lowland relevés on gravelly-stony ground from Hold with Hope were by Sørensen separated as a special group (29) because of the different geographical position. However, the close relation to group 28 is obvious, confirmed by the similarity index (QS = 57), and consequently they are included in the variant (Table 25, anal. 4, 5 and 40).

On almost level ground on a plateau, 25 m a.s.l. at Zackenberg most of the surface is covered by a *Carex rupestris-Salix arctica-Dryas* heath. Almost all ground is covered by cryptogams, yet tiny frostboils with e.g. *Juncus biglumis, Phippsia algida,* and *Cochlearia groenlandica* occur. The soil is acid, indicated by *Hierochlöe alpina.* A relevé (Table 25, anal. 10) is considered a representative of the variant.

6.2.3. Dry, graminoid Dryas heaths and fell-fields

According to Sørensen (s.a.) "Group K represents Saxifragion Cotyledonis Nordhagen 1943 p. 570. Here it includes slightly acidophilous Associations on dry slopes, often with only a thin soil layer covering the bedrock. They present the most resplendent profusion of flowers within the area. Found on (gneissic) bedrock and basalt, missing in the zone of calcareous sediments. Likewise, they seem to be missing at the outer coast, most likely for climatic reasons. The most characteristic species are *Campanula rotundifolia* (= *gieseckiana*), *C. uniflora*, and besides *Saxifraga nivalis*. Further, *Draba daurica* (= *glabella*), *Poa glauca, Hierochlöe alpina.* Includes three different Associations: Ass. r (= K32) with a dense grass mat of *Elyna bellardii* (= *Kobresia myosuroides*), besides with *Carex capillaris*. Forms the transition to Kobresieto-Dryadion. Kap Hedlund, gneissic bedrock. Ass. s (= K33-34) with dwarfshrub vegetation of *Betula nana*, and with *Pyrola grandiflora*. Kap Hedlund, gneissic bedrock. Physiognomically corresponding to Betuleto-Arctostaphyletum alpinae on calcareous ground. Ass. t (= K35) with *Arnica alpina, Draba nivalis, D. fladnizensis*. Traill Ø, basalt, mainly at bird cliffs. Moderately ornithocoprophilous". As, however, Saxifragion cotyledonis consists of clearly chasmophytic associations, these vegetation types are included in Dryadion integrifoliae.

Group K32 is closely related to J31 (QS = 53), undoubtedly both units of Carici-Dryadetum integrifoliae, characteristic of less dry ground than caricetosum rupestri. Groups K33-35 will be treated below under Veronico-Poion glaucae.

Variant of *Carex capillaris* var. nov. (J31)
Five species rich relevés from level or only slightly sloping sites on Traill Ø are dominated by *Dryas, Carex rupestris, C. bigelowii, C. misandra,* and *C. capillaris*. The fairly dry soil is sandy-stony with only a thin humic layer. pH 5.8-6.3 (3 anal.). *Woodsia glabella*, generally occurring only sporadically in Greenland, is character species. Because of its few occurrences pH and conductivity has only been measured at 11 sites with this species (Fig. 23). A typical relevé (Tables 27-28, anal. 134) is from a "dry *Dryas* heath", 200 m a.s.l. pH 6.2. In one relevé only: *Draba lactea, Equisetum variegatum, Kobresia simpliciuscula, Melandrium apetalum, Saxifraga aizoides, Tofieldia coccinea, Cetraria nivalis, Cyrtomnium hymenophylloides, Orthothecium intricatum, Stereocaulon alpinum*, and in the typical relevé: *Polytrichastrum sexangulare* (50%) and *Preissia quadrata* (30%).

Variant of *Hierochloë alpina* var. nov. (K32)
Seven relevés from Kap Hedlund are from level or sloping ground with only slight indications of solifluction. Among the three from 600 m a.s.l., a typical relevé (Tables 27-28, anal. 252) is termed " The most frequent, almost level grass formation at these elevations". Characteristic species is *Hierochloë alpina*, differential species are *Campanula uniflora, Kobresia myosuroides, Saxifraga cernua*, and *Polytrichum juniperinum*. The first two mentioned are acidophilous (Fig. 23), confirmed by pH of 6 soils: 4.6-6.1.

Variant	J31			K32		
No. of relevés			*3*			*7*
F% (F) or average F% (av.)	F	av.		F	av.	
Analysis no.	134			252		
Distichium capillaceum	90	83	*3*	20	39	*6*
Cladonia pyxidata	20	27	*3*	50	51	*6*
Ditrichum flexicaule	50	63	*3*	40	40	*4*
Tortella fragilis		30	*2*	40	29	*5*
Pohlia cruda	90	30	*1*	20	26	*4*
Bryoerythrophyllum recurvirostre	30	17	*2*	30	20	*3*
Isopterygiopsis pulchella	10	20	*3*	10	6	*2*
Amphidium lapponicum		23	*1*		17	*3*
Myurella julacea	10	13	*2*		6	*2*
Aulacomnium turgidum		7	*1*	70	20	*2*
Sanionia uncinatus		3	*1*	50	11	*2*
Encalypta rhabdocarpa	10	7	*2*		1	*1*
Polytrichum strictum		17	*1*	70	10	*1*
Brachythecium groenlandicum	20	7	*1*		6	*1*
Hypnum bambergeri		43	*2*			
Polytrichastrum alpinum	100	37	*2*			
Fissidens osmundoides		7	*2*			
Polytrichum juniperinum					34	*4*
Oncophorus wahlenbergeri				20	6	*2*
Nostoc sp.					6	*2*
Blepharostoma trichophyllum				10	4	*2*
No. of species, range	13	12-16		16	4-16	
No. of species, average		14			11	
Summa F%, range	520	490-650		520	170-650	
Summa F%, average		547			434	

Table 28. Ass. Carici-Dryadetum integrifoliae var. of *Carex capillaris* (J31) and var. of *Hierochloë alpina* (K32), cryptogams.

In one relevé only: *Arnica angustifolia, Calamagrostis purpurascens, Carex supina, Melandrium affine/triflorum, Taraxacum arcticum, Encalypta brevicollis, Hypnum revolutum, Peltigera rufescens, Stegonia latifolia, Stereocaulon paschale, Tortula ruralis,* and in the type relevé: *Cassiope tetragona* (5%), *Dicranum spadiceum* (60%), *Polytrichum piliferum* and *Anastrophyllum minutum* (10%).

7. All. Veronico-Poion glaucae Nordh. 1943

7.1. Middle-arctic, open and very open, graminoid heaths and fell-fields on dry soil

Ass. Cerastio-Festucetum brachyphyllae Daniëls & Fredskild ass. nov. (Z14-15, K34)

The character species are *Festuca brachyphylla, Cerastium arcticum*, and *Draba arctica*. In the 73°N area *Festuca brachyphylla* seems to be common in this association only, assuming that all (or most) *Festuca's* which in Koenigio-Saginetum intermediae by Sørensen were termed *F. brachyphylla* actually are *F. hyperborea*.

Typical variant var. nov. (Z14)

In the lowland of the Zackenberg valley the ridges and slopes with a very thin snow cover have a very open phanerogam and moss cover but many lichens, epigaeic as well as epilithic. Indications of erosion are clear. Only two dwarfshrubs grow here: *Dryas* sp. and *Salix arctica* (Fredskild & Bay 1993, Table 2, anal. 22, 27, 28, 31, anal. 117). A very open *Kobresia myosuroides-Carex rupestris-Festuca brachyphylla-Armeria scabra* vegetation is closely related (Fredskild 1996 anal. 117)

The type relevé of the association (Table 29, anal. 22) is from a 5° S-facing slope, c. 50 m a.s.l. The soil is a stiff, cracked clay (Aug. 8) often with salt efflorescences. Three relevés, 25 m a.s.l., are from level, more gravelly ground. In one relevé only: *Juncus biglumis, Luzula arctica, Potentilla hyparctica*, and in the type relevé: *Alopecurus alpinus* (20%).

Soil characteristica for *Festuca brachyphylla* (Fig. 24), and pH and conductivity for some characteristic and faithful species (Fig. 25) are given.

Variant of *Carex nardina* var. nov. (Z15)

Differential species are *Carex nardina* and *Arenaria pseudofrigida*, whereas *Kobresia myosuroides* is found in only one of 40 circles. The four relevés are from level or SW-facing ground in the lowland at Zackenberg, 36-210 m a.s.l. (Fredskild & Bay 1993, Table 2, anal. 21, 32, 35, 36). A typical relevé (Table 29, anal. 35) is from level ground on top of a little, sandy-gravelly ridge with large stones. Many lichens, mostly epilithic, but also *Bryoria chalybeiformis, Cetraria* sp., and *Thamnolia* sp. In one relevé only: *Draba nivalis* and *D. subcapitata*. E.S. Hansen (Fredskild & al.

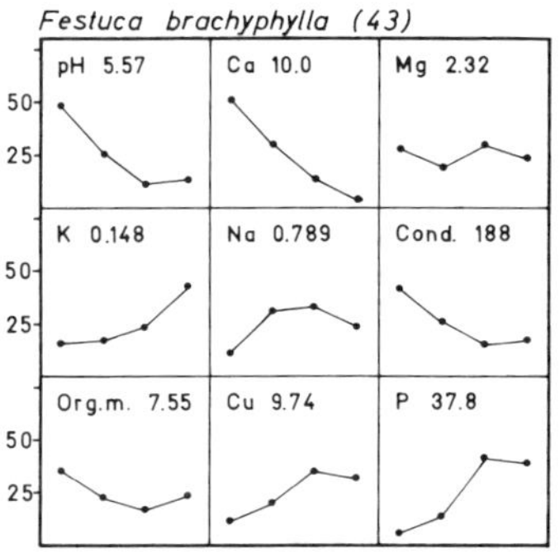

Fig. 24. Soil characteristica for *Festuca brachyphylla*.

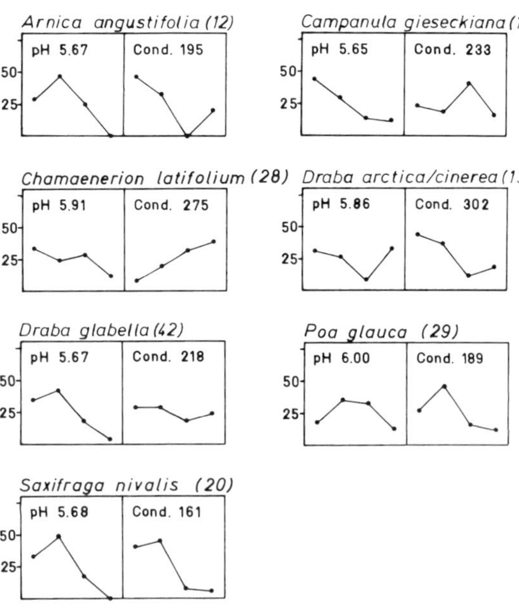

Fig. 25. pH and conductivity for seven species common in two or three of the units of ass. Cerastio-Festucetum brachyphyllae.

Association	Z14														
Variant/community	Z14			Z15			K34			L39		M40		
Area	Z			Z			H			EY		H		
Elevation, m	25-100			36-210			50			80-400		5-50		
No. of relevés			*4*			*4*					*4*			
F% (F) or average F% (av.)	F	av.		F	av.		F	F	F	av.		F	F	F
Analysis no.	22			35			74	76	270			58	78	80
Carex rupestris	80	40	*4*	50	50	*4*	80	35	100	100	*4*	25	45	50
Dryas octopetala/sp.	70	83	*4*	10	58	*4*	55	25	35	41	*4*	50	10	15
Salix arctica	80	55	*4*	+	8	*4*	20	30		5	*2*	60	90	95
Melandrium triflorum/affine	20	18	*3*	+	8	*4*	20	5		41	*4*		20	25
Poa glauca	30	10	*4*	10	8	*3*	25			16	*3*	50	100	85
Cerastium arcticum	40	38	*4*	20	15	*3*	30	10	70	6	*1*		5	
Festuca brachyphylla/sp.	60	38	*4*	+	8	2	15	10		1	*1*	5	25	30
Draba arctica/cinerea	10	20	2	20	13	*3*	10	20	30	36	*4*	5	20	35
Equisetum arvense	30	8	2											
Luzula confusa		18	2		5	*1*	5	10	5					
Poa arctica	10	20	*3*		+	*1*		5						
Hierochloë alpina		8	2		3	*1*	5	5						
Saxifraga oppositifolia		13	2		15	*3*	20			56	*4*			
Silene acaulis		8	2		3	*1*	40	40	35	4	2			
Saxifraga nivalis	10	3	*1*		3	*1*	90		55	4	2			
Saxifraga cernua		3	*1*		5	2	100	50	35			5	35	
Kobresia myosuroides	100	68	*4*		3	*1*				9	*1*		5	
Papaver radicatum		3	*3*	10	3	*1*				13	*1*		10	5
Potentilla hookeriana/nivea	20	13	*3*		+	*1*						15	35	
Polygonum viviparum	10	10	2		8	*1*				8	2			
Minuartia rubella	20	20	2							83	*4*			
Carex nardina				90	68	*4*	15	30		79	*4*			
Arenaria pseudofrigida				+	2	*3*				26	2			
Carex supina							40		100					
Rumex acetosella							30		10					
Chamaenerion latifolium							5			23	2			
Draba glabella							55	20	35	61	*4*			5
Lesquerella arctica										41	*4*	5	5	
No of species, range	16	14-16		13	11-15		20	15	12	13-16		9	13	10
No. of species, average		16			13					14				
Summa F%, range	610	380-640		230	220-420		670	355	600	515-760		220	405	350
Summa F%, average		554			285					628				

Table 29. Ass. Cerastio-Festucetum brachyphyllae (Z14) and related communities, phanerogams.

1995, Table 1, anal. 2) has analyzed the lichens on the site of another of the relevés. Dominating here are *Ochrolechia frigida, Cladonia pocillum, Thamnolia vermicularis, Cetraria muricata, C. nivalis, Physconia muscigena*, and *Alectoria nigricans*. No less than 30 lichens were found here.

Variant of *Draba glabella* var. nov. (K34)
Three relevés (Tables 29-30, anal. 74, 76, 270) from 50 m a.s.l. on Kap Hedlund are open, with very few dwarfshrubs, mainly Dryas. Lichens are frequent. In the 73°N area *Draba arctica/cinerea* has its main occurrence in this community and on raised marine clay in the *Poa glauca* comm. (M40) on the same peninsula. pH of the soil 5.3, 4.9 and 5.4 resp. In one relevé only: *Campanula gieseckiana* (anal. 270: 90%), *Pyrola grandiflora* (anal. 76: 60%), *Schistidium apocarpum* (anal. 74, 10%). Lichens not included in Table 30: *Stereocaulon* sp. 90, 30 and 100%, "other lichens" 100, 0 and 30% respectively in the three relevés.

Draba glabella-Lesquerella arctica community (L39)
Four relevés from Ella Ø and Ymer Ø, 80-400 m a.s.l., differ in the frequent *Carex nardina, Minuartia rubella*, and *Draba glabella. Lesquerella arctica* occur in all relevés, indicating a transitional position to the var. of *Lesquerella arctica* of Arabido holboellii-Caricetum supinae (L38). The cryptogam cover is sparse, as emphasized in the field notes to two of the relevés: cryptogams in only 7 and 4 circles respectively. Cryptogam analyses were made in only two relevés (Table 30), from the one

Association	Z14								
Association/community	K34			L39		K33			K35
No. of relevés								*8*	
F% (F) or average F% (av.)	F	F	F	F	F	F	av.		F
Analysis no.	74	76	270	89	290	70			209
Cladonia pyxidata		90	90	10	42	50	36	*6*	80
Ditrichum flexicaule		10		90	28	50	30	*5*	50
Bryoerythrophyllum recurvirostre			60	40	42		11	*3*	10
Encalypta rhabdocarpa		10	20	30	57		6	*2*	30
Polytrichum piliferum	20	100	30						
Physconia muscigena		10		10					50
Hypnum revolutum	30	20		20		80	36	*6*	10
Pohlia cruda			10	20		10	28	*5*	20
Polytrichum juniperinum	30					40	19	*3*	20
Polytrichastrum alpinum			10				38	*4*	
Cetraria nivalis	10	10					1	*1*	
Encalypta brevicollis	30						3	*1*	
Saelania glaucescens	20						1	*1*	
Racomitrium canescens	40	50							10
Distichium capillaceum				100	71		33	*5*	10
Tortella fragilis				70	42	10	28	*5*	30
Myurella julacea				30			8	*3*	10
Tortula ruralis				90		10	10	*3*	50
Aulacomnium turgidum						10	26	*4*	
Isopterygiopsis pulchella							5	*4*	
Ceratodon purpureus							14	*2*	
Brachythecium groenlandicum						10	5	*2*	
Stegonia latifolia							4	*2*	
Peltigera rufescens						30	14	*4*	20
Amphidium lapponicum							6	*2*	10
No. of species, range	11	10	8	12	8	14	6-16		18
No. of species, average							12		
Summa F%, range	390	420	350	560	371	400	220-650		450
Summa F%, average							464		

Table 30. Ass. Cerastio-Festucetum brachyphyllae (Z14) and related communities, cryptogams.

(anal. 290) in only 7 samples. In one relevé only: *Betula nana, Minuartia stricta, Poa alpina*, and in anal. 290 *Encalypta alpina*. Soil characteristica for *Minuartia rubella* are given in Fig. 26.

Poa glauca community (M40)

Poa glauca is dominant in three relevés from active, clayey solifluction soil on raised marine beds at Kap Hedlund, differing from the otherwise acid soils on the peninsula by the more basic soil: pH in the relevés 6.9, 7.1 and 5.5, resp. On a visit on April 12, 1932 the two first sites (Table 29) were free of snow, the third almost so. In anal. 80: *Poa pratensis* ssp. *alpigena* 5%. Sørensen (s.a.) mentions that on Traill Ø corresponding vegetations are found under similar conditions, but no analyses are available.

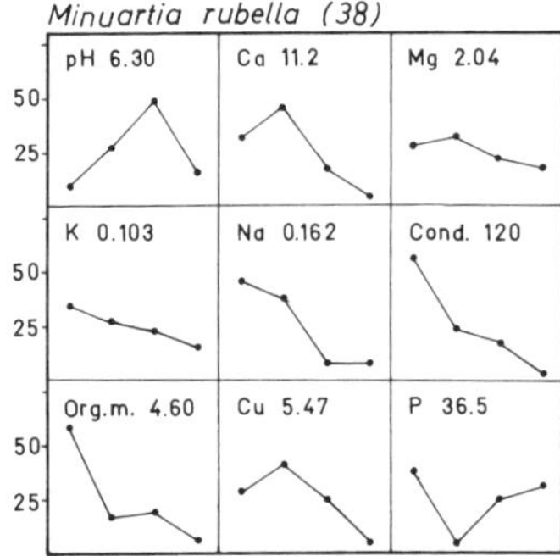

Fig. 26. Soil characteristica for *Minuartia rubella*

Pyrola grandiflora-Betula nana community (K33)

Nine lowland relevés on level or slightly sloping ground on Kap Hedlund are from fairly open vegetations with a high species diversity (Tables 30-31). The cryptogam cover is often small in the relevés below 100 m a.s.l., thus in one relevé moss was found in only three of ten circles. However, *Hypnum revolutum* is a character species of the community, occurring in 6 of 8 relevés (Table 30). Differential species against Cerastio-Festucetum brachyphyllae: *Betula nana, Pyrola grandiflora, Vaccinium uliginosum*, and, partly, *Draba glabella*. The soil profiles are only weakly developed, with pH of the topsoil 5.0-6.1 (8 anal.), subsoil 4.6-6.1 (6 anal.). In one relevé 300 m a.s.l. with a "dry *Betula-Hierochlöe* vegetation with a thick moss layer", dominated by *Aulacomnium turgidum, Polytrichum alpinum*, and *Pohlia nutans*, pH of the topsoil is 5.0, of the subsoil 4.6.

A typical relevé (Tables 30-31, anal. 70) is from a "*Vaccinium* formation, only little solifluction, south exposed, very thin morr layer 1-4 cm, below this the whitish-grey solifluction soil", 40 m a.s.l. In one relevé only: *Draba arctica/cinerea, Luzula nivalis, Aulacomnium palustre, Encalypta alpina, Pohlia nutans, Stereocaulon paschale*, and in the type relevé *Equisetum arvense* (5%), *Sanionia uncinatus* (20%), and *Stereocaulon* sp. (10%). In the 73°N area *Orthotrichum speciosum* and *Polytrichastrum hyperboreum* were only found in one relevé of this community. With its higher cover of dwarfshrubs the community is transitional to the real dwarfshrub heaths.

Community	K33			K35		
Area	H			T		
Elevation, m	20-300			400-650		
No. of relevés			*9*			
F% (F) or average F% (av.)	F	av.		F	F	F
Analysis no.	70			167	168	209
Salix arctica	90	68	*9*	5	10	65
Cerastium arcticum	30	24	*9*	40	75	25
Luzula confusa	5	9	*9*	10	50	20
Carex rupestris	100	79	*8*	75		95
Polygonum viviparum	55	52	*9*			90
Festuca brachyphylla s.l.	50	34	*7*	20	5	30
Poa glauca	15	19	*7*	50	70	20
Saxifraga cernua	15	12	*7*	5	100	75
Saxifraga oppositifolia	55	26	*5*	15	100	10
Dryas octopetala	30	30	*6*		5	65
Silene acaulis	30	18	*5*		80	5
Melandrium triflorum/affine		2	*4*	5	5	
Campanula gieseckiana		7	*3*		75	100
Arnica angustifolia		8	*2*	20	50	95
Kobresia myosuroides		20	*4*			15
Carex nardina		3	*2*	50	95	
Minuartia rubella		3	*2*	70	35	
Draba fladnizensis	15	4	*2*		15	85
Trisetum spicatum	5	1	*2*		100	
Papaver radicatum		1	*1*	20	30	
Calamagrostis purpurascens		2	*1*	5		
Saxifraga caespitosa		1	*1*	5		
Oxyria digyna		1	*1*		5	
Betula nana	95	89	*9*			
Poa arctica	15	43	*7*			
Draba glabella	70	28	*7*			
Pyrola grandiflora	50	49	*6*			
Hierochloë alpina		40	*6*			
Vaccinium microphyllum	100	33	*6*			
Pedicularis hirsuta	10	8	*5*			
Carex bigelowii	5	6	*4*			
Poa pratensis ssp. *alpigena*	60	18	*3*			
Pedicularia lapponica	5	2	*3*			
Potentilla hookeriana/nivea				100	40	70
Campanula uniflora				85	55	30
Draba bellii				50	25	
Taraxacum phymatocarpum				15	50	
Chamanerion latifolium				35	5	5
Draba nivalis				20	10	
No of species	23	13-24		24	30	21
No. of species, average		18				
Summa F%, range	910	530-1035		745	1445	975
Summa F%, average		756				

Table 31. *Pyrola grandiflora-Betula nana* community (K33) and *Potentilla-Campanula uniflora* community (K35), phanerogams.

Potentilla-Campanula uniflora community (K35)

Three relevés (Tables 30-31, anal. 167, 168, 209) on Traill Ø, 400-650 m a.s.l. close to bird cliffs, are moderately ornithocoprophilous. *Arnica angustifolia, Potentilla nivea/hookeriana*, and *Campanula uniflora* are characteristic species. pH is 6.6-6.7 (2 anal.). The phytosociological position is uncertain. Not included in anal. 167: *Erigeron eriocephalus* (20%), *Woodsia glabella* (10%), in anal. 168: *Taraxacum arcticum* (100%), *Draba subcapitata* (75%), *Juncus biglumis* (35%), *Minuartia biflora* (45%), *Poa alpina* (5%), in anal. 209: *Euphrasia arctica* (55%), *Carex capillaris* (5%). Cryptogam analyses only in anal. 209. An early snowbed on Ole Rømer Land (74°08´N), 450 m a.s.l. close to the ice cap is dominated by *Arnica angustifolia, Campanula gieseckiana, C. uniflora*, and *Festuca vivipara*, and with i.a. *Euphrasia frigida, Poa glauca*, and *Trisetum spicatum*, which indicate affinity to this community (Schwarzenbach 1961, p. 136).

7.2. Middle-arctic very open, xerophilous fell-field vegetations

According to Sørensen (s.a.) Group L, consisting of subgroups L36-39 includes "the most xerophilous vegetation type of the area, bound to dry, windswept ridges and summits with sand and gravel, without snowcover during winter and with no traces of solifluction.....Characterized by *Carex supina* ssp. *spaniocarpa* and *Calamagrostis purpurascens*. These species have a western distribution, not found in Scandinavia, where a corresponding vegetation type may not be found. Associations poor in species on gneissic bedrock and basalt (36-37), more species rich on sediments (38-39). Possibly representatives of a special Alliance".

Common to the relevés of the group are: *Carex nardina, C. rupestris, Dryas* sp., and *Kobresia myosuroides*, characteristic species of a major part of the associations of Dryadion integrifoliae, whereas other frequent graminoids and forbs of that alliance, e.g. *Carex bigelowii, C. capillaris, C. misandra, Cerastium arcticum, Luzula confusa, Pedicularis flammea, P. hirsuta*, and *Silene acaulis* are missing.

However, the two character species of the group, *Carex supina* and *Calamagrostis purpurascens*, are all missing in group 39, which consequently has been included in Cerastio-Festucetum brachyphyllae (Z14).

Böcher (1954) in discussing the vegetational complexes in Southwest Greenland describes "The *Artemisia borealis-Calamagrostis purpurascens-Arctostaphylos uva-ursi* Complex. Sub-low arctic continental xerophytic grasslands and dwarf shrub vegetation as well as associated willow scrubs". Ecogeographical guiding species characterizing the complex are: 1) continental sub-low arctic xerophytes: *Carex supina* ssp. *spaniocarpa, Roegneria violacea, Halimolobus mollis, Artemisia borealis, Antennaria affinis*, american races of *Arctostaphylos uva-ursi*, 2) species with a widely ranging arctic-continental distribution: *Calamagrostis purpurascens, Potentilla chamissonis, Erigeron compositus*, (*Melandrium triflorum*), and 3) species associated with the ultrabasic-halobous type: *Braya linearis, B. humilis*, (*Draba lanceolata*), *Gentiana detonsa* var. *groenlandica, Puccinellia deschampsioides*. Climatic indicator species: *Kobresia myosuroides, Rhododendron lapponicum, Dryas integrifolia, Betula nana, Arnica angustifolia, Draba aurea, Primula stricta*.

Böcher divides the complex into three vegetational types, which may be equivalent to alliances (Daniëls 1994): a) *Carex supina spaniocarpa-Potentilla chamissonis* Type, b) *Puccinellia deschampsioides-Gentiana detonsa* Type, and c) *Arctostaphylos uva-ursi-Betula nana* Type. The first mentioned is divided into three "Sociation Groups", which more or less correspond to associations. Among these are *Carex supina spaniocarpa* sociations, *Calamagrostis purpurascens* soc., *Potentilla chamissonis* soc., and *Kobresia myosuroides* soc.

Provisionally, Böchers *Carex supina spaniocarpa-Potentilla chamissonis* Type might be raised to an Alliance: Calamagrostion purpurascentis. Dierssen (1996) treats the unit on the association level: Arabido holboellii-Caricetum supinae. However, according to the rules the publication is invalid as no type relevé is given. The Greenland distribution of this association is low-middle arctic, continental, in W.Greenland found in the inland between Godthåbsfjord (64°N) and Svartenhuk (72°N), in East Greenland, especially northwards somewhat depauperate, from Scoresby Sund (70°N) to Skærfjorden (77½°N). The latter marks the North limit of *Carex supina*, and also the limit of the more frequent occurrences of *Calamagrostis purpurascens*, which further north is very rare, found only in the most continental parts of North Greenland (Bay 1992).

Ass. Arabido holboellii-Caricetum supinae Daniëls & Fredskild ass. nov. (L37-38, Z16)
In W.Greenland the association is found

Subassociation	L37								
Subassociation/variant	L37			L38			Z16		
Area	HT			EY			Z		
Elevation, m	50-600			100-450			75-100		
No. of analyses			*9*			*5*			
F% (F) or average F% (av.)	F	av.		F	av.		F	F	F
Analysis no.	79			113			24	25	26
Carex supina	40	50	*9*	100	38	*4*	90	90	30
Carex rupestris	15	49	*7*	100	100	*5*		90	70
*Dryas octopetala/*sp.	25	45	*9*	85	49	*4*			30
Potentilla hookeriana/nivea	15	10	*2*	80	29	*5*	70	100	70
Kobresia myosuroides		6	*5*	10	58	*4*		10	
Calamagrostis purpurascens	25	23	*4*	70	49	*4*		40	
Poa glauca	10	4	*3*		2	*1*	40	30	40
Lesquerella arctica	5	1	*1*	85	41	*4*	+		
Carex nardina	5	36	*9*	10	55	*5*			
Polygonum viviparum		10	*3*	55	78	*5*			
Saxifraga oppositifolia		8	*3*	15	35	*4*			
Chamaenerion latifolium		2	*2*		25	*2*			
Arenaria pseudofrigida		1	*1*	25	14	*2*			
Draba fladnizensis		2	*1*	20	11	*2*			
Silene acaulis		3	*2*		2	*1*			
Salix arctica	5	2	*3*					10	40
Euphrasia frigida					19	*2*			
Saxifraga cernua				5	13	*2*			
Gentiana detonsa					13	*2*			
Braya linearis					8	*2*			
Melandrium triflorum/affine				60	19	*3*	40	80	50
Minuartia rubella				45	11	*2*	20		
Draba glabella				20	4	*1*			10
Hierochloë alpina					1	*1*			30
Draba arctica							40	40	70
Festuca brachyphylla							20	20	40
Cerastium arcticum							10	10	20
Arnica angustifolia								20	80
Saxifraga nivalis							10	10	
Papaver radicatum							10	+	
No. of species, range	9	4-10		16	12-16		13	14	16
No. of species, average		7			14				
Summa F%, range	145	115-390		785	580-785		360	550	680
Summa F%, average		254			688				

Table 32. Ass. Arabido holboellii-Caricetum supinae, phanerogams.

on south facing slopes around the head of Søndre Strømfjord/Kangerlussuaq on very dry sites, dominated by *Carex supina*, and with *Calamagrostis purpurascens* in five of the analyses (Böcher 1954, Table 20, anal. 1-10), and on hardly as dry sites, dominated by *Calamagrostis purpurascens* with fewer but constant *Carex supina* (l.c. anal. 13-18). Cryptogams poorly developed. Anal. 15 from a 35° SE slope, 75 m a.s.l., is considered the type relevé.

Typical variant var. nov. (L37)

Seven lowland relevés (50-250 m a.s.l.) and one 600 m a.s.l. on Kap Hedlund are from extremely barren, level or only slightly sloping, stony ground. In the field notes is mentioned for several relevés that no moss and virtually no lichens occur, and no cryptogam analyses are at hand. pH 5.4-6.7 (6 anal.). A typical relevé (Table 32, anal. 79) is from a very windswept, stony site, 50 m a.s.l. In one relevé only: *Hierochlöe alpina* and *Luzula confusa*.

A ninth relevé from a stony scree on the basalt plateau 475 m a.s.l. on Traill Ø was by Sørensen because of the different area separated as subgroup 36 consisting of this relevé only. It is included in the table. Not included is *Campanula uniflora* (10%). "Virtually neither moss nor lichens observed" according to the field notes. The association is represented in Ole Rømer Land close to the ice cap (74°N), but no analyses are at hand (Schwarzenbach 1960, Fredskild & al. 1992).

For most species only a limited number of pH and conductivity analyses are at hand (Fig. 27). Judging from this *Carex supina* does not occur on soil with pH above 7. However, in Böcher (1963) ten measurements range between 5.7 and 8.0.

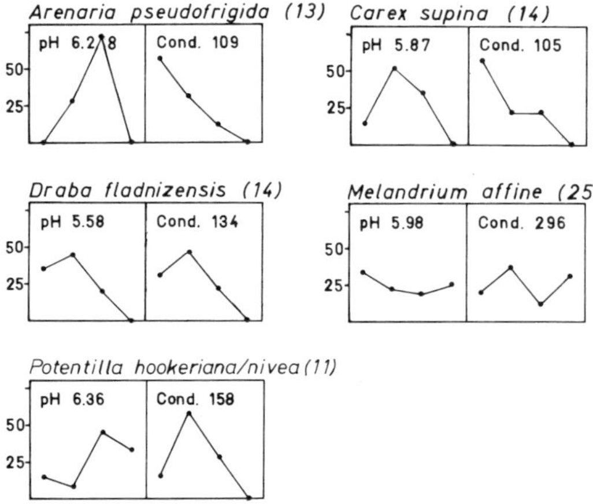

Fig. 27. Soil characteristica for five species common in ass. Arabido holboellii-Caricetum supinae.

The seven analyses on sites with *Calamagrostis purpurascens* (not given here) show the same picture, but Böcher (l.c.) gives pH 7.9 and 8.0 for two sites with the species. No moss analyses available.

Variant of *Lesquerella arctica* var. nov. (L38)
Five relevés from Ella Ø and Ymer Ø, 100-450 m a.s.l., are more species diverse. Differential taxa: *Potentilla nivea* s.l., *Lesquerella arctica, Melandrium affine/triflorum*. The soil is neutral (pH 6.7-7.5, 5 anal.). The occurrence of *Lesquerella arctica* indicate the close affinity to two units of Carici-Dryadetum integrifoliae subass. caricetosum rupestri also found only on Ella Ø and Ymer Ø, viz. the typical variant (J26, QS = 49) and the variant of *Carex glacialis* (J27, QS = 46). A typical relevé (Tables 32-33, anal. 113) is from a south facing slope, 250 m a.s.l., on a slightly elevated part of the ground, otherwise with another of the relevés of the typical variant (L37) from which it differs in the frequent *Calamagrostis purpurascens* and *Potentilla nivea*, and the fewer *Saxifraga cernua* and *S. oppositifolia*. pH is 7.0.

In one relevé only: *Braya humilis, Campanula gieseckiana, Draba nivalis, Luzula confusa, Pedicularis flammea, Encalypta longicollis, Hymenostylium recurvirostrum, Hypnum revolutum, Physconia muscigena,* and *Stegonia latifolia*.

Variant of *Draba arctica* var. nov. (Z16)
Differential species are *Draba arctica, Festuca brachyphylla, Cerastium arcticum*. The three lowland relevés from the Zackenberg valley are from steep slopes (Fredskild & Bay 1993, Table 2). Anal. 25 is forming a long, but only c. 3 m wide, horizontal belt in the uppermost part of the south facing slope of a large moraine. Downslope it is replaced by a only 1-1½ m wide belt with the vegetation of anal. 26. The inclination increases from c. 10° in the uppermost part of the broad belt to c. 35° in the lower part of the narrow belt. Downslope the latter is replaced by en early snowfree *Salix arctica* belt. Anal. 24 is from a 20-30° southwest facing slope of the same moraine, forming a belt between the top and a snowbed vegetation. *Melandrium triflorum*, not *M. affine*, was found in the three relevés. Not in Table 32: anal. 24: *Vaccinium uliginosum* (10%), anal. 25: *Poa arctica* (40%), and in anal. 26: *Stellaria longipes* (80%), *Festuca rubra*, and *Trisetum spicatum* (10%).

Variant	L38		
No. of relevés			*5*
F% (F) or average F% (av.)	F	av.	
Analysis no.	113		
Distichium capillaceum	20	54	*5*
Tortella fragilis	30	48	*5*
Bryoerythrophyllum recurvirostre	50	38	*4*
Encalypta rhabdocarpa	70	24	*4*
Tortula ruralis	20	12	*4*
Amphidium lapponicum	70	32	*3*
Ditrichum flexicaule	40	20	*3*
Cladonia pyxidata		18	*3*
No. of species, range	9	7-11	
No. of species, average		9	
Summa F%, range	370	170-390	
Summa F%, average		310	

Table 33. Ass. Arabido holboellii-Caricetum supinae var. of *Lesquerella arctica*, cryptogams.

Area	HH		
F% (F)	F	F	F
Analysis no.	32	52	53
Festuca brachyphylla s.l.	35	64	40
Taraxacum phymatocarpum	50	4	25
Poa abbreviata	30	12	25
Saxifraga oppositifolia	10	18	20
Poa glauca	10	2	30
Papaver radicatum	5	2	10
Carex maritima	100	24	
Puccinellia angustata	55		40
Cerastium arcticum		14	15
Draba bellii		12	10
Minuartia rubella		4	15
Melandrium triflorum/affine		2	15
Carex nardina		6	10
Potentilla pulchella		4	10
Polygonum viviparum		2	10
Stellaria longipes s.l.	5		5
Salix arctica		2	5
Potentilla rubricaulis	5	2	
Poa hartzii			55
Draba subcapitata			20
Draba cinerea			15
Draba adamsii	10		
No. of species	14	19	20
Summa F%	330	180	380

Table 34. *Taraxacum phymatocarpum-Poa abbreviata* community (N41), phanerogams.

7.3. Very open vegetation on solifluction soil

Taraxacum phymatocarpum-Poa abbreviata community (N41)

Phytosociologically three lowland relevés (Table 34) from Hold with Hope stand quite isolated, as illustrated by the QS of their group (N41) against the other 40 groups. In only two cases it is higher than 30, viz. 38 to group 35, and 32 to group 40. The syntaxonomical position is unclear, but provisionally it is placed under Veronico-Poion glaucae. Many of its species have a high-arctic distribution with their East Greenland S-limit at Scoresby Sund: *Taraxacum phymatocarpum, Puccinellia angustata, Draba bellii, Potentilla pulchella, P. rubricaulis, Braya purpurascens, Draba micropetala (incl. in D. adamsii), Luzula arctica, Poa hartzii*, and *Colpodium vahlianum*, or slightly more to the south on the Blosseville Coast or at Kangerdlugssuaq: *Arenaria pseudofrigida, Draba subcapitata*, and *Poa abbreviata*. However, the dominating role of *Festuca brachyphylla*, only common northwards to 78°N, might indicate that the community is characteristic of the Middle Arctic Tundra Zone sensu Elvebakk (1985). In describing the dry clayey flats and raised beaches 74½°-79°N Sørensen (in Seidenfaden and Sørensen 1937) mentions many of the above species as characteristic for such sites, and even if it is not mentioned in the field notes his relevés most likely are from raised marine clay. In one relevé only (anal. 32): *Braya purpurascens, Luzula arctica*, and *Colpodium vahlianum* 5%, in anal. 52: *Arenaria pseudofrigida, Campanula uniflora*, and *Potentilla nivea*: 2%, and in anal. 53: *Draba nivalis* 5%.

8. All. Puccinellion phryganodis Hadac 1946 em. Hofmann 1969

8.1. Middle-arctic halophytic vegetation

Because of the eroding effect of the ice saltmarshes are only found on protected sites, usually in deltas. At the outlet of the Zackenberg river a 1 x 1½ km area of tidal flats and raised beaches is partly covered by vegetations of the association Koenigio-Saginetum intermediae, partly by true saltmarshes, and by some intermediate vegetations. The low-arctic littoral vegetations of Southeast Greenland are described in detail by de Molenaar (1974), whereas the generally very poor middle-arctic ones are shortly mentioned by Sørensen in Seidenfaden & Sørensen (1937) and Bay (1992).

Ass. Puccinellietum phryganodis Hadac 1946

The circumpolar *Puccinellia phryganodes* s.l. includes several lower taxa, e.g. the diploid *P. phryganodes* s.str. in the Alaska-Chukotka area, the tetraploid *P. vilfoidea* in Spitsbergen, Scandinavia and Siberia, and the triploid *P. neoarctica* in Greenland. Consequently, Hadac (1989) renamed Puccinellietum phryganodis to Puccinellietum vilfoideae, based on *Phippsia vilfoidea* Löve and Löve. If this view as to the taxonomy is accepted, the correct name of the Greenland association is Puccinellietum neoarcticae.

As described in de Molenaar (1974) the representatives of this association often consist of only one species, viz. *Puccinellia phryganodes*, sometimes accompanied by *Stellaria humifusa*. In the former delta at Zackenberg large, outer parts of the tidal flats, and a narrow belt along the tidal channels in the inner part are covered by a "monoculture" of *Puccinellia phryganodes* without mosses (Table 35, anal. 124), surrounded by a belt with many *Stellaria humifusa* (anal. 125).

In the Zackenberg delta Puccinellietum phryganodis is mostly replaced on slightly higher level by vegetations dominated by *Carex ursina*, with many *Puccinellia phryganodes* and scattered *Stellaria humifusa* (anal. 126). Locally, along the tidal channels in the inner part a narrow belt of either pure *Carex subspathacea* or *C. subspathacea-Puccinellia phryganodes* stands is found between the *Puccinellia phryganodes* zone and the *Carex ursina* zone. Here, *Stellaria humifusa* is generally more common, with scattered *Carex subspathacea* (anal. 130). Such *Carex ursina-Puccinellia phryganodes-Stellaria humifusa* belts may on higher level be replaced by a mossy belt with some *Koenigia islandica* (anal. 129).

Sørensen, in Seidenfaden & Sørensen (1937) mentions three species "which often occur as facultative halophytes at the margins of the shore lagoons, where, together with *Stellaria humifusa*, they form a kind of shore snow-patch vegetation, viz. *Sagina intermedia, Phippsia algida*, and *Saxifraga rivularis*, the latter, however, probably represented by a special "salt ecotype", distinguished morphologically by a lower and more compact growth, more succulent stems and leaves, and a lighter green colour ..... than the normal form on salt-free soil". Beyond doubt this is *S. hyperborea*, which is common in such transitional vegetations, e.g. in a *Saxifraga hyperborea-Phippsia algida-Stellaria humifusa-Puccinellia* snowbed (anal. 127), found as a belt between a *Carex supina-Puccinellia phryganodes* vegetation and a several metre high erosion slope with a *Saxifraga hyperborea-Luzula confusa-Ranunculus pygmaeus* snowbed and, further up, a herb-slope with i.a. *Arnica angustifolia, Erigeron humilis*, and *E. eriocephalus*.

Ass. Caricetum subspathaceae Hadac 1946

Frequently, a *Carex subspathacea* belt with many *Stellaria humifusa* and *Koenigia islandica* (anal. 128) is found between

Caricetum ursinae and the true non-halophytic vegetations, e.g. moist *Salix arctica-Arctagrostis latifolia-Eriophorum triste* heaths.

Carex subspathacea is far from restricted to the littoral areas. Frequently, around lakes and ponds frequented by geese, the border is covered by a one to several metre wide, thick, manured green moss layer in which it may be the only phanerogam. The two examples (anal. 107, 135) are from 80-90 m a.s.l. In Herb.C. many collections are from the lowland below 100 m, but even to 300 m it has been collected at a "goose lake" in Northeast Greenland.

Analysis no.	124		125		130		126		129		128		127		107		135	
Puccinellia phryganodes	*30*	100	*30*	100	*24*	100	*16*	80	*22*	100	*8*	50	*10*	80				
Stellaria humifusa			*26*	100	*26*	100	*4*	40	*20*	100	*12*	70	*23*	90				
Carex ursina			*1*	10	*24*	100	29	100	*15*	80								
Carex subspathacea					*3*	20			*17*	90	*27*	100			*30*	100	*30*	100
Koenigia islandica									*6*	40	*12*	70						
Saxifraga hyperborea											*3*	20	*21*	100				
Phippsia algida													*18*	100	2	10		
Carex glareosa											*3*	20						
Mosses							*10*	60	*15*	70	*24*	100	*27*	100	*30*	100	*30*	100

Table 35. All. Puccinellion phryganodis at Zackenberg. Numbers in italics denote the total score (max. 30), see methods p. 10.

9. Concluding remarks

In Daniëls (1994) an account to classify the vegetation of Greenland is presented. It includes a first survey of higher syntaxa from the subarctic-low arctic part, but because of lack of publications the middle and high arctic could not be considered. As to Northeast Greenland the first and only attempt at classifying the plant communities was Seidenfaden & Sørensen (1937), who arranged them into 18 "main ecosystems and vegetation types of Northeast Greenland 74°30'-79°N.lat." (l.c. p. 110). In a table the occurrence of 130 phanerogam species in the 18 types is given. In his undated summary Sørensen only compares the vegetations between 72° and 74°N with those in Scandinavia, described in Nordhagen (1943). Thus, his groups A-N, characterized by a similarity quotient (QS) 40-50, are paralleled to Nordhagen's Alliances, "yet in between only with difficulty as an absolute parallel in the vegetation units does not exist for the two regions: as to plant sociology Northeast Greenland is not a piece of the alpine Scandinavia yet single Associations may be refound fairly unchanged". Generally, the use of Sørensens similarity coefficient has resulted in well defined phytosociological units. The very few exceptions, mostly of units consisting of one relevé only, e.g. his subgroup J30.1 mentioned under Saliceto-Cassiopetum tetragonae, are caused by the equal "weighing" of species, whether highly ubiquitous (e.g. *Salix arctica* and *Saxifraga oppositifolia*) or highly selective (e.g. *Potentilla crantzii* and *Carex supina*).

Three of Sørensens groups had no Scandinavian parallel, and he suggested (Sørensen s.a.) three new alliances, viz. Junco-Koenigio islandicae, Arctagrostideon latifoliae, and an unnamed, based on the very open xerophilous fell-field vegetations characterized by *Carex supina* and *Calamagrostis purpurascens*. These groups are included in three of the five alliances in which all Northeast Greenland vegetation types have been grouped.

As to morphology and ecology many of the lower syntaxa of Northeast Greenland are paralleled to those of Svalbard, which are summarized by Elvebakk (1994). However, phytogeographically the two areas are quite different, and only two associations of Puccinellion phryganodis are in common. Further to these, the vegetation types are grouped in 11 associations, of which the 7 are new, 9 new subassociations, and many lower phytosociological units. A survey is given for 43 units, each based on at least four relevés, for 134 phanerogam taxa in the synoptical table (Table 36) worked out by F. Daniëls. A survey of mosses and some lichens in 24 of these units is given in Table 37.

Circumneutral *Cassiope tetragona* communities within the alliance Kobresio-Dryadion Nordhagen 1936 make up the zonal vegetation of the middle arctic tundra zone (Elvebakk 1985) which according to him includes the area 70°-78°N in Northeast Greenland (l.c. map Fig. 3). Bay (1997) divides the middle arctic tundra zone of Northeast Greenland into a coastal part with *Cassiope* heath, fellfield and *Salix arctica* snowbed as the dominating vegetations and *Carex ursina* as the indicator species, and a continental part, with *Cassiope* heath, *Betula nana* heath (south of 75°N), *Salix arctica* snowbed, *Carex stans* grassland, and *Carex stans* fen dominating, and *Ranunculus nivalis*, *Draba alpina*, and *Epilobium arcticum* as indicator species. According to him, the limits of the zone are 69° and 79½°N, illustrated by the fact than many of the associations found in the 72°-74½°N area have their northernmost, although often somewhat depauperate occurrences on Lambert Land. Unfortunately, only six relevés are available from the area 74½°-79°N but summary descriptions, sometimes illustrated by profiles of slopes with lists of the most frequent species occurring in the different zones, related to the duration of the snow cover, are given in Bay & Fredskild (1990, 1991).

10. Acknowledgements

As it appears from the introduction the basic material for this paper originates from Th. Sørensens field work 1931-35, in the 1960'es supplemented by the analyses of many soil samples made at the Department of Plant Ecology, and by the moss determinations by K. Holmen. In the mid-1970'es Mogens Køie and other colleagues at the Botanical Institutes, University of Copenhagen, continued the working up of the material. Thanks to all these persons the final treatment, supplemented by own investigations, was made possible when in 1993 the material was handed over to me.

During this final stage I have received much help from my colleagues at the Botanical Museum. Gert Steen Mogensen has helped with the nomenclature of the mosses in bringing Holmens nomenclature up-to-date. Eric Steen Hansen has determined lichen samples from Zackenberg, and Christian Bay is thanked for the joint fieldwork through many summers, three of which in Northeast Greenland.

Based on early drafts Fred Daniëls, University of Münster, Germany, made a great effort in bringing my phytosociological units in conformity with practise in the Braun-Blanquet system, and I have followed his suggestions as to the level and naming of the units. Further, he worked out the synoptical table 36. I am greatly indebted to him for his giant work with the present paper.

Vegetation type	Z	A	A	D	Z	D	B	Z	B	Z	C	Z	F	F	F	F	Z	Z	E	E	E	E	F	J	J
	1	1	2	9	4	10	5	3	4	2	7	5	18	20	15	17	9	7	11	12	13	14	16	25	24
Number of relevés	6	5	8	5	5	4	16	11	14	7	7	7	6	8	7	7	4	6	12	12	6	19	5	14	39
Ch/D Saxifrago-Ranunculion																									
Minuartia biflora	5	4	5	5	3	.	4	2	2	.	2	1	.	.	.	.	.	.	.	.	.	.	.	.	.
Ranunculus pygmaeus	4	1	5	5	4	.	.	.	.	.	.	.	.	.	.	.	.	.	.	.	.	.	.	.	.
Ranunculus sulphureus	2	.	2	3	.	2	2	.	1	.	3	.	2	2	.	.	.	.	.	.	.	.	.	.	.
Taraxacum arcticum	2	.	.	1	3	.	3	3	2	.	1	1	.	3	.	.	.	.	.	.	.	.	.	.	.
Trisetum spicatum	**5**	**5**	**5**	2	5	.	3	4	3	.	.	.	.	.	.	.	.	.	.	.	.	.	.	.	.
Oxyria digyna	**4**	**4**	**5**	.	5	.	4	4	5	.	.	.	.	2	.	.	.	.	.	.	.	.	.	.	.
Erigeron humilis	**4**	**3**	**5**	3	1	.	1	.	1	.	.	.	.	.	.	.	.	.	.	.	.	.	.	.	.
Festuca rubra	**3**	**4**	**5**	2	1	.	1	.	.	.	.	.	.	.	.	.	.	.	.	.	.	.	.	.	.
Antennaria canescens	.	**4**	**4**	.	.	.	.	.	.	.	.	.	.	.	.	.	.	.	.	.	.	.	.	.	.
Poa alpina	.	**3**	**5**	.	.	.	.	.	.	.	.	.	.	.	.	.	.	.	.	.	.	.	.	.	.
Potentilla crantzii	.	**2**	**5**	.	.	.	.	.	.	.	.	.	.	.	.	.	.	.	.	.	.	.	.	.	.
Taraxacum brachyceras	.	**3**	**3**	.	.	.	.	.	.	.	.	.	.	.	.	.	.	.	.	.	.	.	.	.	.
Antennaria porsildii	.	**1**	**5**	.	.	.	.	.	.	.	.	.	.	.	.	.	.	.	.	.	.	.	.	.	.
Luzula spicata	.	.	**4**	.	.	.	.	.	.	.	.	.	.	.	.	.	.	.	.	.	.	.	.	.	.
Thalictrum alpinum	.	.	**4**	.	.	.	.	.	.	.	.	.	.	.	.	.	.	.	1	.	.	.	.	.	.
Sibbaldia procumbens	.	.	**2**	.	.	.	.	.	.	.	.	.	.	.	.	.	.	.	.	.	.	.	.	.	.
Draba crassifolia	.	.	**2**	.	.	.	.	.	.	.	.	.	.	.	.	.	.	.	.	.	.	.	.	.	.
Gentiana tenella	.	.	2	.	.	.	.	.	.	.	.	.	.	.	.	.	.	.	.	.	.	.	.	.	.
Salix herbacea	.	.	2	**5**	**5**	3	.	.	.	.	.	.	.	.	.	.	.	.	.	.	.	.	.	.	.
Carex lachenalii	.	.	3	3	.	.	.	.	.	.	.	.	.	.	.	.	.	.	.	.	.	.	.	.	.
Saxifraga tenuis	.	.	.	.	.	.	5	4	5	4	5	4	3	2	.	.	.	.	.	.	.	.	.	.	.
Sagina intermedia	.	.	.	.	.	.	5	3	4	2	5	5	2	1	.	.	.	.	.	.	.	.	.	.	.
Draba adamsii	.	.	.	.	.	.	2	.	4	.	2	2	.	.	.	.	.	.	.	.	.	.	.	.	.
Draba alpina	.	.	2	.	.	.	2	.	2	.	2	1	.	.	.	.	.	.	.	.	.	.	.	.	.
Ranunculus nivalis	.	.	.	.	.	.	1	2	1	.	.	.	.	.	.	.	.	.	.	.	.	.	.	.	.
Cerastium regelii	.	.	.	.	.	.	.	.	2	.	1	1	.	.	.	.	.	.	.	.	.	.	.	.	.
Saxifraga hyperborea/rivularis	.	.	.	.	.	.	1	1	**4**	**5**	.	.	.	.	.	.	.	.	.	.	.	.	.	.	.
Phippsia algida	.	.	.	.	.	.	.	2	**4**	**5**	1	.	.	.	.	.	.	.	.	.	.	.	.	.	.
Koenigia islandica	.	.	.	2	.	.	.	.	.	.	**5**	**3**	5	.	.	.	.	.	.	.	.	.	.	.	.
Festuca hyperborea	.	.	.	.	.	.	.	.	.	.	**5**	**4**	.	.	.	.	.	.	.	.	.	.	.	.	.
Deschampsia brevifolia	.	.	.	.	.	.	.	.	.	.	**3**	**2**	.	.	.	.	.	.	.	.	.	.	.	.	.
Saxifraga hirculus	.	.	.	.	.	.	.	.	.	.	**1**	**4**	.	.	.	.	.	.	.	.	.	.	.	.	.
Saxifraga platysepala	.	.	.	.	.	.	.	.	.	.	**2**	**3**	.	.	.	.	.	.	.	.	.	.	.	.	.
Ranunculus glacialis	.	.	.	.	.	.	.	.	.	.	**3**	.	.	.	.	.	.	.	.	.	.	.	.	.	.
Ch/D Caricion atrofusco-saxatilis																									
Pedicularis flammea	.	1	1	.	.	.	.	.	.	.	.	2	.	.	3	2	2	2	5	4	4	5	5	4	4
Juncus triglumis	.	.	.	.	.	.	.	.	.	.	.	1	5	1	1	3	.	.	2	4	4	2	3	1	1
Juncus castaneus	.	.	.	.	.	.	.	.	.	.	2	1	5	.	2	2	5	5	1	1	.	1	4	.	.
Carex saxatilis	.	.	.	.	.	.	.	.	.	.	.	.	5	1	2	5	5	4	.	4	2	2	4	.	.
Carex atrofusca	.	.	.	.	.	.	.	.	.	.	.	1	1	.	.	2	.	.	1	4	.	2	4	.	.
Carex parallela	.	.	.	.	.	.	.	.	.	.	.	.	1	.	3	3	.	.	1	4	2	4	1	1	1
Tofieldia pusilla	.	.	.	.	.	.	.	.	.	.	.	.	.	.	2	.	.	.	2	1	.	3	4	1	3
Carex scirpoidea	.	2	4	.	.	.	.	.	.	.	.	.	.	.	1	1	.	.	3	3	4	3	2	1	4
Eutrema edwardsii	.	.	.	.	.	.	.	.	.	.	2	.	.	.	.	.	.	.	1	1	.	3	.	1	1
Carex pseudolagopina	.	.	.	.	.	.	.	.	.	.	.	.	.	.	.	2	2	.	.	1	.	1	.	.	.
Carex rariflora	.	.	.	.	.	.	.	.	.	.	.	.	.	.	.	2	.	.	.	1	.	2	3	.	.
Eriophorum callitrix	.	.	.	.	.	.	.	.	.	.	.	.	.	.	2	4	.	.	.	2	.	3	4	.	.
Eriophorum scheuchzeri	.	.	.	.	.	.	.	.	.	.	.	.	2	.	1	5	5	.	.	.	.	.	2	.	.
Armeria scabra	.	.	.	.	.	.	.	.	.	.	.	.	.	.	.	.	.	.	.	1	.	.	.	2	1
Arctagrostis latifolia	.	.	.	.	.	.	.	.	.	.	1	1	**5**	**4**	**5**	**5**	**5**	**5**	1	3	1	2	1	1	1
Eriophorum triste	.	.	.	.	.	.	.	.	.	.	1	3	**5**	**5**	**5**	**5**	**4**	**5**	3	5	4	5	5	2	1
Kobresia simpliciuscula	.	.	.	.	.	.	.	.	.	.	.	4	.	.	.	3	.	.	**5**	**5**	**3**	**5**	3	3	2
Saxifraga nathorstii	.	.	.	.	.	.	.	.	.	.	.	.	.	.	.	.	.	.	**2**	**5**	**3**	**4**	.	.	1
Saxifraga aizoides	.	.	.	.	.	.	.	.	.	.	.	.	.	.	.	.	.	.	**5**	**5**	**5**	**4**	.	2	2
Braya purpurascens	.	.	.	.	.	.	.	.	.	.	.	.	.	.	.	.	.	.	**3**	**4**	.	**2**	.	.	1
Minuartia stricta	.	2	.	.	.	.	2	.	.	.	.	.	.	.	.	.	.	.	**2**	**1**	**5**	**2**	.	1	1
Rhododendron lapponicum	.	.	.	.	.	.	.	.	.	.	.	.	.	.	.	.	.	.	.	.	.	1	**5**	**5**	**2**
Tofieldia coccinea	.	.	.	.	.	.	.	.	.	.	.	.	.	.	4	.	.	.	.	.	.	.	**4**	**5**	**1**
Ch/D Dryadion integrifoliae																									
Pedicularis hirsuta	2	.	.	3	1	2	4	2	1	.	1	4	4	4	3	2	.	.	1	2	5	2	1	1	2
Carex misandra	.	.	.	.	.	.	4	2	1	.	5	3	4	5	3	1	2	1	5	5	5	4	5	3	4
Carex rupestris	1	1	.	.	.	.	.	.	.	.	.	2	1	3	4	.	.	4	5	3	3	4	2	5	5
Carex nardina	1	3	.	.	.	.	4	1	.	.	1	1	.	.	.	.	.	.	3	.	3	1	.	3	4
Kobresia myosuroides	.	.	.	.	.	.	.	1	.	.	.	3	.	.	1	.	.	.	5	2	.	2	2	4	3
Papaver radicatum	1	1	.	.	.	.	4	.	1	.	2	.	.	.	.	.	.	.	.	.	.	.	.	1	1
Hierochlöe alpina	2	.	.	.	.	.	.	.	.	.	.	.	.	.	.	.	.	2	.	.	.	.	.	2	.
Lesquerella arctica	.	.	.	.	.	.	.	.	.	.	.	.	.	.	.	.	.	.	2	.	.	.	.	.	.
Chamaenerion latifolium	.	.	.	.	.	.	.	.	.	.	.	.	.	.	.	.	.	.	3	.	.	.	.	1	1
Cassiope tetragona	.	.	.	.	3	.	1	1	.	.	.	.	.	3	4	.	.	.	1	1	.	3	2	3	5
Huperzia selago	.	.	.	.	.	.	.	.	.	.	.	.	.	2	.	.	.	.	.	.	.	.	.	.	.
Carex bigelowii	2	.	5	3	2	4	.	.	.	.	1	1	5	3	3	1	4	5	.	.	5	.	4	4	.
Carex capillaris	.	.	.	.	.	.	.	.	.	.	.	1	.	.	4	.	.	5	1	2	1	2	4	5	2

Table 36(a). Character, differential, and other species of the four NE.Greenland alliances.

Z	Z	G	J	J	J	J	K	J	J	J	Z	Z	L	K	I	L	L	Vegetation type
13	12	21	30	30	30	31	32	26	27	28	14	15	39	33	23	37	38	
5	6	15	11	5	16	5	7	10	15	8	4	4	4	9	11	9	5	Number of relevés
																		Ch/D Saxifrago-Ranunculion
.	.	.	.	.	.	.	.	.	.	.	.	.	.	.	1	.	.	*Minuartia biflora*
.	.	.	.	.	.	.	.	.	.	.	.	.	.	.	.	.	.	*Ranunculus pygmaeus*
.	.	.	.	.	.	.	.	.	.	.	.	.	.	.	.	.	.	*Ranunculus sulphureus*
.	.	.	.	.	.	.	.	.	.	.	.	.	.	.	.	.	.	*Taraxacum arcticum*
.	.	.	.	.	.	.	.	.	.	.	.	.	.	2	.	.	.	*Trisetum spicatum*
1	1	2	1	.	3	.	.	.	.	.	.	.	.	1	.	.	.	*Oxyria digyna*
.	.	.	.	.	.	.	.	.	.	.	.	.	.	.	.	.	.	*Erigeron humilis*
.	.	.	.	.	.	.	.	.	.	.	.	.	.	.	1	.	.	*Festuca rubra*
.	.	.	.	.	.	.	.	.	.	.	.	.	.	.	.	.	.	*Antennaria canescens*
.	.	.	.	.	.	.	.	.	.	.	.	.	.	.	2	.	.	*Poa alpina*
.	.	.	.	.	.	.	.	.	.	.	.	.	.	.	.	.	.	*Potentilla crantzii*
.	.	.	.	.	.	.	.	.	.	.	.	.	.	.	.	.	.	*Taraxacum brachyceras*
.	.	.	.	.	.	.	.	.	.	.	.	.	.	.	.	.	.	*Antennaria porsildii*
.	.	.	.	.	.	.	.	.	.	.	.	.	.	.	.	.	.	*Luzula spicata*
.	.	.	.	.	.	.	.	.	.	.	.	.	.	.	.	.	.	*Thalictrum alpinum*
.	.	.	.	.	.	.	.	.	.	.	.	.	.	.	.	.	.	*Sibbaldia procumbens*
.	.	.	.	.	.	.	.	.	.	.	.	.	.	.	.	.	.	*Draba crassifolia*
.	.	.	.	.	.	.	.	.	.	.	.	.	.	.	.	.	.	*Gentiana tenella*
.	.	.	.	.	.	.	.	.	.	.	.	.	.	.	.	.	.	*Salix herbacea*
.	.	.	.	.	.	.	.	.	.	.	.	.	.	.	.	.	.	*Carex lachenalii*
.	.	.	.	.	.	.	.	.	.	.	.	.	.	.	.	.	.	*Saxifraga tenuis*
.	.	.	.	.	.	.	.	.	.	.	.	.	.	.	.	.	.	*Sagina intermedia*
.	.	.	.	.	.	.	.	.	.	.	.	.	.	.	.	.	.	*Draba adamsii*
.	.	.	.	.	.	.	.	.	.	.	.	.	.	.	.	.	.	*Draba alpina*
.	.	.	.	.	.	.	.	.	.	.	.	.	.	.	.	.	.	*Ranunculus nivalis*
.	.	.	.	.	.	.	.	.	.	.	.	.	.	.	.	.	.	*Cerastium regelii*
.	.	.	.	.	.	.	.	.	.	.	.	.	.	.	.	.	.	*Saxifraga hyperborea/rivularis*
.	.	.	.	.	.	.	.	.	.	.	.	.	.	.	.	.	.	*Phippsia algida*
.	.	.	.	.	.	.	.	.	.	.	.	.	.	.	.	.	.	*Koenigia islandica*
.	.	.	.	.	.	.	.	.	.	.	.	.	.	.	.	.	.	*Festuca hyperborea*
.	.	.	.	.	.	.	.	.	.	.	.	.	.	.	.	.	.	*Deschampsia brevifolia*
.	.	.	.	.	.	.	.	.	.	.	.	.	.	.	.	.	.	*Saxifraga hirculus*
.	.	.	.	.	.	.	.	.	.	.	.	.	.	.	.	.	.	*Saxifraga platysepala*
.	.	.	.	.	.	.	.	.	.	.	.	.	.	.	.	.	.	*Ranunculus glacialis*
																		Ch/D Caricion atrofusco-saxatilis
.	.	.	.	.	.	1	2	2	1	.	.	.	.	.	.	.	.	*Pedicularis flammea*
.	.	.	.	.	.	.	.	.	.	1	.	.	.	.	.	.	.	*Juncus triglumis*
.	.	.	.	.	.	.	.	.	.	.	.	.	.	.	.	.	.	*Juncus castaneus*
.	.	.	.	.	.	.	.	.	.	.	.	.	.	.	.	.	.	*Carex saxatilis*
.	.	.	.	.	.	.	.	.	.	.	.	.	.	.	.	.	.	*Carex atrofusca*
.	.	.	.	.	.	.	.	.	.	.	.	.	.	.	.	.	.	*Carex parallela*
.	.	.	.	.	.	.	.	.	.	.	.	.	.	.	.	.	.	*Tofieldia pusilla*
.	.	.	2	.	1	2	1	1	.	.	.	.	.	.	.	.	.	*Carex scirpoidea*
.	.	.	.	.	.	.	.	.	.	.	.	.	.	.	.	.	.	*Eutrema edwardsii*
.	.	.	.	.	.	.	.	.	.	.	.	.	.	.	.	.	.	*Carex pseudolagopina*
.	.	.	.	.	.	.	.	.	.	.	.	.	.	.	.	.	.	*Carex rariflora*
.	.	.	.	.	.	.	.	.	.	.	.	.	.	.	.	.	.	*Eriophorum callitrix*
.	.	.	.	.	.	.	.	.	.	.	.	.	.	.	.	.	.	*Eriophorum scheuchzeri*
.	.	.	.	.	.	.	.	.	.	.	.	.	.	.	.	.	.	*Armeria scabra*
2	5	.	1	3	2	.	2	.	.	.	.	.	.	.	.	.	.	*Arctagrostis latifolia*
3	.	1	.	.	.	.	3	.	.	.	.	.	.	.	.	.	.	*Eriophorum triste*
.	.	.	.	.	.	.	.	1	.	.	.	.	.	.	.	.	.	*Kobresia simpliciuscula*
.	.	.	.	.	.	.	.	.	.	.	.	.	.	.	.	.	.	*Saxifraga nathorstii*
.	.	.	.	.	.	.	.	.	.	.	.	.	.	.	.	.	.	*Saxifraga aizoides*
.	.	.	.	.	.	.	.	2	.	.	.	.	.	.	.	.	.	*Braya purpurascens*
.	.	.	.	.	.	.	.	.	.	.	.	.	.	.	.	.	.	*Minuartia stricta*
.	.	.	.	.	.	.	.	.	.	.	.	.	.	.	.	.	.	*Rhododendron lapponicum*
2	.	2	.	.	.	.	.	.	.	.	.	.	.	.	.	.	.	*Tofieldia coccinea*
																		Ch/D Dryadion integrifoliae
.	3	3	1	3	3	3	2	1	1	2	.	.	.	3	.	.	.	*Pedicularis hirsuta*
.	5	1	4	2	4	5	2	2	.	3	.	.	.	.	.	.	.	*Carex misandra*
1	4	2	5	4	3	5	5	5	5	5	5	5	5	5	5	4	5	*Carex rupestris*
.	.	.	4	2	3	3	1	4	5	5	.	5	5	2	5	5	5	*Carex nardina*
.	.	1	2	1	.	1	5	5	1	1	5	2	2	3	.	3	4	*Kobresia myosuroides*
.	4	1	2	3	2	2	2	.	.	.	4	2	2	1	.	.	.	*Papaver radicatum*
2	5	3	.	4	.	.	4	.	.	.	3	2	.	4	.	.	1	*Hierochlöe alpina*
.	.	.	.	.	.	.	.	3	4	.	.	.	5	.	1	1	4	*Lesquerella arctica*
.	.	1	.	.	.	1	2	1	1	1	.	.	3	.	.	2	2	*Chamaenerion latifolium*
5	**5**	**5**	**5**	**5**	**5**	.	.	.	.	.	.	.	.	.	.	.	.	*Cassiope tetragona*
.	**2**	**1**	**2**	.	**3**	.	.	.	.	.	.	.	.	.	.	.	.	*Huperzia selago*
3	.	4	5	3	4	5	4	.	.	.	.	.	.	3	.	.	.	*Carex bigelowii*
.	.	2	2	2	1	5	5	.	.	1	.	.	.	.	.	.	.	*Carex capillaris*

Table 36(b).

Vegetation type	Z	A	A	D	Z	D	B	Z	B	Z	C	Z	F	F	F	F	Z	Z	E	E	E	E	F	J	J
	1	1	2	9	4	10	5	3	4	2	7	5	18	20	15	17	9	7	11	12	13	14	16	25	24
Number of relevés	6	5	8	5	5	4	16	11	14	7	7	7	6	8	7	7	4	6	12	12	6	19	5	14	39
Ch/D Veronico-Poion glaucae																									
Poa glauca	1	2	.	.	.	.	1	.	1	.	.	3	.	.	.	.	.	.	.	.	.	.	.	.	.
Melandrium triflorum/affine	1	2	.	.	.	.	.	.	.	.	.	.	.	.	.	.	.	.	2	1	.	.	.	1	1
Potentilla hookeriana/nivea	.	.	.	.	.	.	.	.	.	.	.	.	.	.	.	.	.	.	.	.	.	.	.	.	.
Draba arctica/cinerea	.	.	.	.	.	.	.	.	.	.	.	.	.	.	.	.	.	.	.	.	.	.	.	.	.
Arenaria pseudofrigida	.	.	.	.	.	.	.	.	.	.	.	.	.	.	.	.	.	.	.	.	.	.	.	.	1
Betula nana	.	.	.	.	.	.	.	.	.	.	.	.	.	.	5	.	.	.	2	2	.	5	3	2	5
Pyrola grandiflora	.	.	.	.	.	.	.	.	.	.	.	.	.	.	3	.	.	.	.	.	.	.	.	.	2
Calamagrostis purpurascens	.	.	.	.	.	.	.	.	.	.	.	.	.	.	.	.	.	.	.	.	.	.	.	.	.
Carex supina	.	.	.	.	.	.	.	.	.	.	.	.	.	.	.	.	.	.	.	.	.	.	.	1	.
Gentiana detonsa	.	.	.	.	.	.	.	.	.	.	.	.	.	.	.	.	.	.	.	.	.	.	.	.	.
Other species																									
Polygonum viviparum	5	5	5	5	4	5	3	3	2	.	5	5	5	5	5	5	5	5	5	5	5	5	5	5	5
Salix arctica	5	5	5	4	5	5	5	5	5	3	5	5	5	5	5	5	5	5	5	5	5	5	5	4	5
Dryas octopetala/integrifolia	3	5	.	1	.	2	3	2	.	.	2	5	.	5	4	2	3	2	5	5	5	5	4	5	5
Saxifraga oppositifolia	1	5	1	.	.	.	5	2	4	2	4	3	.	5	3	.	.	.	5	3	5	4	2	4	5
Silene acaulis	5	5	5	4	2	.	4	2	3	.	2	4	2	2	5	.	.	.	3	3	5	2	1	4	5
Luzula confusa	4	.	.	5	5	5	5	5	5	5	3	5	3	3	3	.	.	2	.	.	3	.	2	3	1
Saxifraga cernua	4	4	5	2	1	.	5	5	5	1	5	5	3	4	1	.	3	2	.	.	2	.	.	2	.
Poa arctica	5	.	1	3	4	5	2	5	3	.	3	5	1	4	2	.	2	4	.	.	.	.	1	.	.
Juncus biglumis	.	.	2	2	.	.	5	2	3	1	5	5	5	4	3	3	5	5	3	3	5	2	2	2	1
Equisetum arvense	2	1	4	.	.	.	1	1	.	1	5	2	4	4	5	3	5	3	1	3	3	3	3	1	2
Luzula arctica	.	.	.	2	4	2	5	5	3	.	5	5	5	5	5	.	.	2	1	.	3	1	.	2	1
Draba lactea	4	.	2	3	.	.	5	5	4	.	5	5	3	5	2	1	.	.	.	1	4	1	.	1	1
Stellaria longipes s.l.	2	.	.	2	1	5	2	4	3	2	5	5	5	3	.	.	.	2	.	.	.	.	.	.	.
Vaccinium microphyllum	.	.	.	1	.	4	.	.	.	.	.	1	2	2	5	2	.	5	2	1	.	5	5	5	4
Festuca brachyphylla	2	.	.	2	1	.	2	1	2	.	.	.	.	2	.	.	.	3	.	.	.	.	.	.	.
Cerastium arcticum	5	5	5	2	1	.	5	5	5	.	5	4	.	.	.	.	.	.	.	.	.	.	.	.	.
Equisetum variegatum	1	1	3	2	1	.	.	.	.	.	.	2	1	4	1	5	2	.	3	5	3	5	.	.	3
Cardamine bellidifolia	.	.	.	.	.	.	3	2	3	1	3	2	3	4	4	.	.	.	.	.	.	.	.	2	.
Saxifraga foliolosa	.	.	.	.	.	.	1	2	2	3	5	2	5	2	1	1	2	1	.	.	3	.	1	.	.
Minuartia rubella	.	2	.	.	.	.	4	1	2	.	3	3	.	2	.	.	.	.	.	.	.	.	.	.	.
Draba glabella	4	5	2	.	.	.	2	.	.	.	.	.	.	.	.	.	.	.	.	.	.	.	.	1	1
Potentilla hyparctica	5	.	.	5	2	5	2	3	1	.	3	1	3	3	.	.	.	2	.	.	.	.	.	.	.
Alopecurus alpinus	4	.	.	.	4	.	.	3	.	2	5	3	2	.	.	.	3	3	.	.	.	.	.	.	.
Melandrium apetalum	.	.	.	.	.	.	3	.	.	.	4	3	2	.	.	1	.	.	1	1	2	.	.	.	.
Arctostaphylos alpina	.	.	.	.	.	.	.	.	.	.	.	.	.	.	.	.	.	.	.	.	.	1	.	1	2
Empetrum hermaphroditum	.	.	.	.	.	.	.	.	.	.	.	.	.	.	3	.	.	.	.	.	.	.	1	1	1
Pedicularis lapponica	.	.	.	.	.	.	.	.	.	.	.	.	.	.	3	.	.	.	1	.	.	3	.	1	3
Draba fladnizensis	.	.	2	.	.	.	.	.	.	.	.	.	.	.	.	.	.	.	.	.	.	.	.	.	.
Arnica angustifolia	1	1	1	.	.	.	.	.	.	.	.	.	.	.	.	.	.	.	.	.	.	.	.	.	.
Braya linearis	.	.	.	.	.	.	.	.	.	.	.	.	.	.	.	.	.	.	2	2	.	.	.	.	.
Saxifraga nivalis	1	.	1	2	4	2	.	.	.	.	.	2	.	.	.	.	.	.	.	.	.	.	.	.	.
Campanula uniflora	.	.	.	2	.	2	.	.	.	.	.	.	.	.	.	.	.	.	.	.	.	.	.	.	.
Carex maritima	.	.	.	.	.	.	.	.	.	.	3	2	3	.	.	.	.	.	.	2	.	.	.	.	.
Draba subcapitata	.	.	.	.	.	.	2	.	1	.	2	.	.	.	.	.	.	.	.	.	.	.	.	.	.
Euphrasia frigida	.	2	2	.	.	.	.	.	.	.	.	.	.	.	.	.	.	.	2	2	.	.	.	2	.
Poa pratensis ssp. *alpigena*	.	.	3	.	.	.	.	.	.	.	.	.	.	.	.	.	.	.	.	.	1	1	.	2	1
Rumex acetosella	.	.	.	.	.	.	.	.	.	.	.	.	.	.	.	.	.	.	.	.	.	.	.	1	.
Saxifraga caespitosa	.	.	.	.	.	.	1	.	.	.	.	2	.	.	.	.	.	.	.	.	.	.	.	.	.
Woodsia glabella	.	.	.	.	.	.	.	.	.	.	.	.	.	.	.	.	.	.	1	.	.	.	.	1	.
Carex microglochin	.	.	.	.	.	.	.	.	.	.	.	.	.	.	.	.	.	.	.	2	.	.	.	.	.
Carex norvegica	.	.	.	.	.	.	.	.	.	.	.	.	.	.	.	.	.	2	.	.	.	.	.	.	.
Cochlearia groenlandica	.	.	.	.	.	.	.	.	2	.	.	.	.	.	.	.	.	.	.	.	.	.	.	.	.
Colpodium vahlianum	.	.	.	.	.	.	.	.	.	.	3	.	.	.	.	.	.	.	.	.	.	.	.	.	.
Saxifraga hieraciifolia	.	.	.	.	.	.	1	.	.	.	.	.	.	.	.	.	.	.	.	.	.	.	.	.	.
Draba bellii	.	.	.	.	.	.	.	.	.	.	3	.	.	.	.	.	.	.	.	.	.	.	.	.	.
Carex glacialis	.	.	.	.	.	.	.	.	.	.	.	.	.	.	.	.	.	.	.	.	.	.	.	.	1
Draba nivalis	.	.	.	.	.	.	1	.	1	.	.	.	.	.	.	.	.	.	.	.	.	.	.	.	.
Campanula gieseckiana	.	.	.	.	.	.	.	.	.	.	.	.	.	.	.	.	.	.	.	.	.	.	.	.	.
Festuca baffinensis	.	.	.	.	.	.	.	.	.	.	.	2	.	.	.	.	.	.	.	.	.	.	.	.	.
Festuca vivipara	.	.	.	.	.	.	.	1	.	.	.	.	.	.	.	.	.	.	.	.	.	.	.	.	.
Juncus arcticus	.	.	.	.	.	.	.	.	.	.	.	.	.	.	.	.	.	.	.	1	.	.	.	.	.
Luzula wahlenbergii	.	.	.	.	.	.	.	.	.	.	.	.	.	.	.	.	3	.	.	.	.	.	.	.	.
Dupontia psilosantha	.	.	.	.	.	.	.	.	.	.	.	.	.	.	.	.	2	.	.	.	.	.	.	.	.
Carex bicolor	.	.	.	.	.	.	.	.	.	.	.	.	.	.	.	1	.	.	.	1	.	.	.	.	.

Table 36(c).

Z 13	Z 12	G 21	J 30	J 30	J 30	J 31	K 32	J 26	J 27	J 28	Z 14	Z 15	L 39	K 33	I 23	L 37	L 38	Vegetation type
5	6	15	11	5	16	5	7	10	15	8	4	4	4	9	11	9	5	Number of relevés
																		Ch/D Veronico-Poion glaucae
.	.	.	.	2	.	2	2	.	.	.	5	4	4	4	3	2	1	*Poa glauca*
.	.	.	.	.	.	.	.	.	.	.	4	5	5	3	2	.	3	*Melandrium triflorum/affine*
.	.	.	.	.	.	.	.	.	1	.	4	2	.	.	.	2	5	*Potentilla hookeriana/nivea*
.	.	.	.	.	.	.	.	1	1	.	3	4	5	.	1	.	.	*Draba arctica/cinerea*
.	.	.	1	.	1	.	.	1	.	1	.	4	3	.	2	1	2	*Arenaria pseudofrigida*
.	.	4	.	.	.	.	.	1	1	.	.	.	.	**5**	**5**	.	.	*Betula nana*
.	.	5	.	.	.	.	.	1	.	.	.	.	.	**4**	**3**	.	.	*Pyrola grandiflora*
.	.	.	.	.	.	.	.	.	1	.	.	.	.	1	.	**3**	**4**	*Calamagrostis purpurascens*
.	.	.	.	.	.	.	.	.	.	.	.	.	.	.	.	**5**	**4**	*Carex supina*
.	.	.	.	.	.	.	.	.	.	.	.	.	.	.	.	.	**2**	*Gentiana detonsa*
																		Other species
3	5	4	5	5	5	5	5	4	2	4	3	2	3	5	3	2	5	*Polygonum viviparum*
5	5	5	5	5	5	5	5	4	2	5	5	5	3	5	3	2	.	*Salix arctica*
5	5	2	5	5	5	5	5	5	5	5	5	5	5	4	5	5	4	*Dryas octopetala/integrifolia*
.	1	2	3	1	5	5	4	5	5	5	3	4	5	3	5	2	4	*Saxifraga oppositifolia*
.	1	2	4	3	5	5	4	3	4	4	3	2	3	3	4	2	1	*Silene acaulis*
1	1	5	1	5	5	5	5	.	.	2	3	2	.	5	.	.	.	*Luzula confusa*
.	.	1	1	.	3	2	5	1	.	3	2	3	.	4	1	.	2	*Saxifraga cernua*
3	5	4	1	4	2	2	1	.	.	1	4	2	.	4	.	.	.	*Poa arctica*
.	.	.	1	1	1	3	3	.	.	.	.	.	.	.	.	.	.	*Juncus biglumis*
.	1	1	.	2	.	2	1	.	.	.	3	.	.	.	.	.	.	*Equisetum arvense*
1	5	3	2	3	4	2	.	.	.	2	.	.	.	.	.	.	.	*Luzula arctica*
.	1	1	.	2	2	.	.	.	.	.	.	.	.	.	.	.	.	*Draba lactea*
2	5	.	1	4	3	3	.	.	.	.	.	.	.	.	.	.	.	*Stellaria longipes* s.l.
5	5	5	5	5	.	.	.	.	.	.	.	.	.	4	.	.	.	*Vaccinium microphyllum*
.	3	2	1	3	1	.	2	.	.	1	5	3	2	4	.	.	.	*Festuca brachyphylla*
.	.	.	.	.	1	.	2	.	.	2	5	4	2	5	2	.	.	*Cerastium arcticum*
.	.	.	1	1	1	.	.	.	.	.	.	.	.	.	.	.	.	*Equisetum variegatum*
1	3	3	1	1	3	1	2	.	.	.	.	.	.	.	.	.	.	*Cardamine bellidifolia*
.	.	.	.	.	.	.	.	.	.	.	.	.	.	.	.	.	.	*Saxifraga foliolosa*
.	.	.	.	.	.	1	1	.	.	2	3	.	5	2	3	.	2	*Minuartia rubella*
.	.	1	.	.	1	.	3	1	1	.	.	.	5	4	3	.	1	*Draba glabella*
.	.	.	.	.	.	.	2	.	.	1	.	.	.	.	.	.	.	*Potentilla hyparctica*
1	5	.	.	2	.	.	.	.	.	.	.	.	.	.	.	.	.	*Alopecurus alpinus*
.	.	.	.	.	.	.	.	.	.	.	.	.	.	.	.	.	.	*Melandrium apetalum*
.	.	.	.	.	.	.	.	2	1	.	.	.	.	.	3	.	.	*Arctostaphylos alpina*
5	.	2	.	.	.	.	.	.	.	.	.	.	.	.	.	.	.	*Empetrum hermaphroditum*
.	.	3	.	.	.	.	.	.	.	.	.	.	.	2	.	.	.	*Pedicularis lapponica*
.	.	.	.	.	.	2	2	.	.	.	.	.	.	2	.	1	2	*Draba fladnizensis*
.	.	.	.	.	.	.	.	.	.	1	.	.	.	2	.	.	.	*Arnica angustifolia*
.	.	.	.	.	.	.	.	1	1	.	.	.	.	.	.	.	2	*Braya linearis*
.	.	.	.	.	.	.	2	.	.	.	2	2	3	.	2	.	.	*Saxifraga nivalis*
.	.	.	.	.	.	1	4	.	.	.	.	.	.	.	.	.	.	*Campanula uniflora*
.	.	.	.	.	.	.	.	.	.	.	.	.	.	.	.	.	.	*Carex maritima*
.	.	.	.	.	.	.	.	.	.	2	.	.	.	.	.	.	.	*Draba subcapitata*
.	.	.	.	.	.	1	1	1	1	.	.	.	.	.	.	.	2	*Euphrasia frigida*
.	.	.	2	.	1	1	2	.	.	.	.	.	.	2	1	.	.	*Poa pratensis* ssp. *alpigena*
.	.	.	.	.	.	.	2	.	.	.	.	.	.	.	.	.	.	*Rumex acetosella*
.	.	.	.	.	.	.	.	.	.	.	.	.	.	1	1	.	.	*Saxifraga caespitosa*
.	.	.	.	.	.	4	.	1	.	.	.	.	.	.	.	.	.	*Woodsia glabella*
.	.	.	.	.	.	.	.	.	.	.	.	.	.	.	.	.	.	*Carex microglochin*
.	.	.	.	.	.	.	.	.	.	.	.	.	.	.	.	.	.	*Carex norvegica*
.	.	.	.	.	.	.	.	.	.	.	.	.	.	.	.	.	.	*Cochlearia groenlandica*
.	.	.	.	.	.	.	.	.	.	.	.	.	.	.	.	.	.	*Colpodium vahlianum*
.	.	.	.	.	.	.	.	.	.	.	.	.	.	.	.	.	.	*Saxifraga hieraciifolia*
.	.	.	.	.	.	.	.	.	.	.	.	.	.	.	.	.	.	*Draba bellii*
.	.	.	1	.	.	.	.	.	4	.	.	.	.	.	.	.	.	*Carex glacialis*
.	.	.	.	.	.	.	.	.	.	.	.	.	.	.	1	.	.	*Draba nivalis*
.	.	.	.	.	.	.	2	.	.	.	.	.	.	2	.	.	.	*Campanula gieseckiana*
.	.	.	.	.	.	.	.	.	.	.	.	.	.	.	.	.	.	*Festuca baffinensis*
.	.	.	.	.	.	.	.	.	.	.	.	.	.	.	.	.	.	*Festuca vivipara*
.	.	.	.	.	.	.	.	.	.	.	.	.	.	.	.	.	.	*Juncus arcticus*
.	.	.	.	.	.	.	.	.	.	.	.	.	.	.	.	.	.	*Luzula wahlenbergii*
.	.	.	.	.	.	.	.	.	.	.	.	.	.	.	.	.	.	*Dupontia psilosantha*
.	.	.	.	.	.	.	.	.	.	.	.	.	.	.	.	.	.	*Carex bicolor*

Table 36(d).

Vegetation type	A 1	A 2	B 5	B 4	F 20	F 15	F 17	E 11	E 12	E 14	F 16	J 25	J 24	G 21	J 30	J 31	K 32	J 26	J 27	J 28	L 39	K 33	I 23	L 38
Number of relevés	4	3	7	9	7	7	5	10	11	19	5	11	36	15	17	3	3	8	11	3	4	8	8	5
Amphidium lapponicum	.	.	1	1	1	.	.	.	.	1	1	2	1	1	1	1	2	1	1	.	.	1	1	2
Aneura pinguis	.	.	.	.	.	.	2	.	.	1	.	.	1	.	.	.	.	.	.	.	.	.	.	.
Aulacomnium palustre	.	.	.	.	1	1	2	.	.	1	2	.	.	.	1	.	1	.	.	.	.	1	.	.
Aulacomnium turgidum	.	.	.	.	2	3	1	.	.	.	2	2	.	3	1	1	.	.	.	.	.	2	.	.
Bartramia ithyphylla	.	.	1	2	1	.	.	.	.	.	.	1	.	.	1	.	.	.	.	.	.	.	.	.
Blepharostoma trichophyllum	.	.	1	.	1	3	1	1	1	2	2	1	1	2	.	.	1	1	.	.	.	.	.	.
Brachythecium binervulum	.	.	.	.	.	1	.	1	1	1	.	1	2	.	.	1	.	.	1	.	.	.	.	.
Brachythecium groenlandicum	.	.	.	1	.	.	.	.	.	.	.	.	1	1	1	1	1	1	1	.	.	1	.	.
Brachythecium salebrosum	.	.	.	.	1	.	.	.	.	1	.	.	.	.	.	.	.	.	.	.	.	.	.	.
Brachythecium turgidum	.	.	.	.	.	.	.	1	1	1	.	.	1	.	.	.	.	.	.	.	.	.	.	.
Bryoerythrophyllum recurvirostre	2	.	.	1	2	.	.	2	2	2	.	2	2	1	1	1	2	2	3	1	2	2	2	3
Bryum neodamense	.	.	.	.	.	.	.	.	1	1	.	.	.	.	.	.	.	.	.	.	.	.	.	.
Bryum wrightii	.	.	.	.	.	.	.	1	.	1	.	.	.	.	.	.	.	.	.	.	.	.	.	.
Calliergon giganteum	.	.	.	.	.	.	2	.	.	.	.	.	.	.	.	.	.	.	.	.	.	.	.	.
Calliergon sarmentosum	.	.	.	.	.	.	.	.	.	.	2	.	.	.	.	.	.	.	.	.	.	.	.	.
Calliergon trifarium	.	.	1	.	.	.	2	1	1	1	2	1	1	.	.	.	.	.	.	.	.	.	.	.
Campylium stellatum	.	.	.	.	2	2	3	2	2	3	2	1	1	.	.	.	.	1	.	1	.	.	.	.
Campylopus schimperi	.	.	.	.	.	.	.	.	.	.	.	1	.	.	.	.	.	.	.	.	.	.	.	.
Catoscopium nigritum	.	.	.	.	.	.	2	.	2	1	.	.	.	.	.	.	.	.	.	.	.	.	.	.
Ceratodon purpureus	.	.	1	1	.	.	.	.	.	.	.	.	1	1	.	.	.	1	.	.	.	1	1	.
Cesia concinnata	.	.	.	1	.	.	.	.	.	.	.	.	.	.	2	.	.	.	.	.	.	.	.	.
Cetraria delisei	.	.	3	1	.	.	.	.	.	.	1	1	.	.	2	.	.	.	.	2	.	.	.	.
Cetraria nivalis	.	.	.	.	.	.	.	1	.	.	.	1	1	1	2	1	.	.	1	1	.	1	1	.
Cinclidium arcticum	.	.	.	.	.	.	2	.	1	1	.	.	.	.	.	.	.	.	.	.	.	.	.	.
Cinclidium subrotundum	.	.	.	.	.	.	1	.	.	.	2	.	.	.	.	.	.	.	.	.	.	.	.	.
Cirriphyllum cirrhosum	.	.	.	.	.	.	.	.	.	1	.	.	1	.	.	.	.	.	.	.	.	.	.	.
Cladonia pyxidata	1	2	3	2	2	1	.	1	.	1	2	2	2	1	3	1	3	2	2	2	3	3	2	2
Conostomum tetragonum	.	.	.	1	.	.	.	.	.	.	.	.	.	.	1	.	.	.	.	.	.	.	.	.
Cyrtomnium hymenophylloides	.	.	.	.	2	.	2	2	1	1	1	1	1	.	.	1	.	1	.	.	.	.	.	.
Dicranum angustum	.	.	.	.	.	.	1	.	.	.	.	.	.	.	.	.	.	.	.	.	.	.	.	.
Dicranum elongatum	.	.	.	.	.	.	.	.	.	.	.	1	.	1	.	.	.	.	.	.	.	.	.	.
Dicranum fuscescens	.	.	.	.	.	.	.	.	.	.	.	1	.	1	1	.	.	.	.	.	.	.	.	.
Dicranum fusc. var. *congestum*	.	.	.	.	.	.	.	.	.	.	.	.	.	.	1	.	.	.	.	.	.	.	.	.
Dicranum spadiceum	.	.	.	.	1	.	.	.	.	.	.	1	.	2	2	.	1	.	.	.	.	.	.	.
Didymodon asperifolius	.	.	.	.	1	.	.	.	.	.	.	.	1	.	.	.	.	1	.	.	.	.	.	.
Distichium capillaceum	2	2	3	3	3	2	2	3	3	3	2	3	3	1	3	1	3	3	3	2	2	2	2	3
Distichium inclinatum	.	.	.	.	.	.	.	.	1	.	.	.	.	.	.	.	.	1	.	.	.	.	.	.
Ditrichum flexicaule	1	.	2	2	3	2	3	3	3	3	3	3	3	2	3	1	2	3	3	2	2	2	3	2
Drepanocladus aduncus s.l.	.	.	.	.	.	.	.	.	1	1	.	.	1	.	.	.	.	1	.	.	.	.	.	.
Drepanocladus badius	.	.	.	.	1	2	2	.	.	.	2	.	.	.	.	.	.	.	.	.	.	.	.	.
Drepanocladus brevifolius	.	.	.	.	.	.	3	.	1	1	.	1	.	.	.	.	.	.	.	.	.	.	.	.
Drepanocladus intermedius	.	.	.	.	1	.	1	.	1	1	.	.	1	.	.	.	.	.	.	.	.	.	.	.
Drepanocladus revolvens	.	.	.	.	1	1	2	.	.	1	2	.	.	.	.	.	.	.	.	.	.	.	.	.
Encalypta alpina	.	.	.	.	2	.	.	.	1	1	.	.	.	.	.	.	.	1	.	1	1	1	1	.
Encalypta brevicollis	.	.	.	.	.	.	.	.	1	1	.	.	1	.	.	.	1	.	.	.	1	1	.	.
Encalypta longicollis	.	.	.	.	.	.	.	2	.	.	.	1	1	.	.	.	.	2	2	.	.	.	.	1
Encalypta procera	.	.	.	.	.	.	.	1	1	1	.	1	1	.	1	.	.	1	1	.	.	.	.	.
Encalypta rhabdocarpa	.	.	1	.	.	.	.	1	.	1	.	1	1	.	1	1	1	1	1	.	2	1	3	3
Fissidens osmundoides	.	.	1	.	2	.	.	.	.	1	2	.	1	.	1	1	.	.	.	.	.	.	.	.
Hylocomium splendens	.	.	.	.	1	.	.	.	.	.	.	.	.	.	.	.	.	1	.	.	.	.	.	.
Hymenostylium recurvirostrum	.	.	.	.	.	.	.	1	.	1	.	1	.	.	.	.	.	.	.	.	.	.	.	1
Hypnum bambergeri	.	.	1	.	1	.	1	3	3	2	.	2	2	1	1	1	.	2	1	1	.	.	.	.
Hypnum callichroum	.	.	.	.	1	.	.	.	.	.	.	1	.	1	1	.	.	.	.	.	.	.	.	.
Hypnum pratense	.	.	.	.	1	1	2	.	.	.	.	.	.	.	.	.	.	.	.	.	.	.	.	.
Hypnum revolutum	.	.	2	.	.	.	.	.	.	1	.	.	2	2	1	.	1	1	1	.	1	3	1	1
Isopterygiopsis pulchella	.	.	1	.	3	2	.	1	.	1	2	1	1	1	2	1	1	1	.	1	.	2	.	.
Lophozia rutheana	.	.	.	.	.	.	2	.	.	.	2	.	.	.	.	.	.	.	.	.	.	.	.	.
Meesia triquetra	.	.	.	.	.	.	2	.	.	.	.	.	.	.	.	.	.	.	.	.	.	.	.	.
Meesia uliginosa	.	.	.	.	1	1	3	2	2	1	2	.	1	.	1	.	.	.	.	.	.	.	.	.
Mnium thomsonii	.	.	1	.	.	.	.	.	.	1	.	.	1	.	.	.	.	.	.	.	.	.	.	.
Myurella julacea	.	.	1	.	3	1	.	2	1	1	2	2	1	.	1	1	1	2	1	1	1	2	1	.
Myurella tenerrima	.	.	.	.	1	1	.	.	.	.	.	.	.	.	.	.	.	.	.	.	.	.	.	.
Odontoschisma macounii	.	.	.	.	1	1	.	.	.	.	2	.	1	.	1	.	.	.	.	.	.	.	.	.
Oncophorus wahlenbergii	.	.	.	.	1	2	.	.	.	1	2	1	1	2	1	.	1	1	.	.	.	.	.	.
Orthothecium chryseum	.	.	.	.	1	.	2	.	1	2	1	.	1	.	1	.	.	.	.	.	.	.	.	.
Orthothecium intricatum	.	.	.	.	.	.	.	.	1	.	.	.	1	.	.	.	.	.	.	.	.	.	.	.
Peltigera rufescens	.	.	.	.	.	.	.	.	.	.	.	.	1	1	1	.	1	.	.	.	.	2	1	.
Philonotis tomentella	.	.	.	.	2	1	2	.	.	.	.	.	.	.	.	.	.	.	.	.	.	.	.	.
Physcia muscigena	.	.	.	.	.	.	.	1	.	.	.	.	.	.	.	.	.	.	.	.	.	.	.	.

Table 37(a). Mosses and some lichens in vegetation types with at least three relevés.

Vegetation type	A 1	A 2	B 5	B 4	F 20	F 15	F 17	E 11	E 12	E 14	F 16	J 25	J 24	G 21	J 30	J 31	K 32	J 26	J 27	J 28	L 39	K 33	I 23	L 38
Number of relevés	4	3	7	9	7	7	5	10	11	19	5	11	36	15	17	3	3	8	11	3	4	8	8	5
Platydictya jungermannioides	.	.	.	.	.	.	.	1	.	.	.	1	1	.	.	.	.	.	.	.	.	.	1	.
Pohlia cruda	.	.	3	1	2	2	.	.	.	.	1	2	1	2	2	1	2	1	.	1	1	2	1	.
Pohlia nutans	.	.	.	.	1	2	1	.	.	1	.	.	.	2	.	.	.	.	.	.	.	1	.	.
Pohlia obtusifolia	.	.	.	1	.	.	.	.	.	.	.	.	.	.	.	.	.	.	.	.	.	.	.	.
Polytrichastrum alpinum	.	2	1	2	1	1	1	.	.	.	2	1	1	1	1	1	.	.	.	.	.	2	1	.
Polytrichum juniperinum	.	.	2	1	.	.	.	.	.	.	.	1	.	.	2	.	2	.	.	1	.	2	.	.
Polytrichum piliferum	.	.	.	1	.	.	.	.	.	.	.	1	.	1	1	.	1	.	.	.	.	.	.	.
Polytrichum strictum	.	.	1	.	2	1	1	.	.	.	1	1	.	2	1	1	1	.	.	.	.	.	.	.
Preissia quadrata	.	.	.	.	.	.	.	.	.	.	.	.	.	.	.	1	.	1	.	.	.	.	.	.
Psilopium cavifolium	.	.	.	1	.	.	.	.	.	.	.	.	.	.	1	.	.	.	.	.	.	.	.	.
Ptilidium ciliare	.	.	.	.	.	.	.	.	.	.	.	.	.	1	1	.	.	.	.	.	.	.	.	.
Rhacomitrium canescens	.	.	.	.	.	.	.	.	.	.	.	.	.	.	1	.	.	.	.	.	.	.	.	.
Saelania glaucescens	.	.	1	.	.	.	.	.	.	.	.	.	.	.	1	.	.	.	.	.	.	1	.	.
Sanionia uncinatus	.	.	1	1	1	1	1	.	.	1	.	1	1	2	2	1	1	.	.	.	.	1	.	.
Schistidium apocarpum	.	.	.	.	.	.	.	1	1	1	.	.	1	.	1	.	.	1	.	.	.	.	1	.
Scorpidium turgescens	.	.	.	1	1	1	2	1	2	2	.	1	1	.	.	.	.	1	.	.	.	.	.	.
Solorina octospora	.	.	.	.	.	.	.	1	1	1	.	1	1	.	1	.	.	1	1	.	.	.	.	.
Sphenolobus minutus	.	.	.	.	1	1	.	.	.	.	2	1	.	1	1	.	1	.	.	.	.	.	.	.
Stegonia latifolia	1	.	.	.	.	.	.	.	1	.	.	.	.	.	.	.	1	.	1	.	.	1	1	1
Stereocaulon alpinum	.	.	.	.	.	.	.	.	.	.	.	1	.	.	2	1	.	.	.	.	.	.	.	.
Stereocaulon paschale	.	.	2	1	.	.	.	1	.	1	.	1	.	.	.	.	1	2	1	1	.	1	1	.
Tayloria lingulata	.	.	.	.	.	.	1	.	.	.	1	.	.	.	.	.	.	.	.	.	.	.	.	.
Thamnolia vermicularis	.	.	.	.	.	.	.	1	1	.	.	1	1	.	1	.	.	.	1	.	.	.	1	.
Timmia austriaca	.	.	.	.	1	.	.	.	.	.	.	.	.	1	1	.	.	.	.	.	.	.	.	.
Timmia norvegica	.	.	.	.	.	.	1	.	.	.	.	.	.	.	.	.	.	.	.	.	.	.	.	.
Tomenthypnum nitens	.	.	.	.	1	3	2	.	.	2	.	1	1	1	1	.	.	.	.	.	.	.	.	.
Tortella fragilis	.	.	3	.	3	1	2	2	2	2	1	2	3	1	2	1	2	3	2	1	2	2	2	3
Tortula mucronifolia	.	.	.	.	.	.	.	.	.	.	.	.	.	.	.	.	.	1	.	.	.	.	1	.
Tortula norvegica	.	2	.	.	.	.	.	.	.	.	.	.	.	.	.	.	.	.	.	.	.	.	.	.
Tortula ruralis	.	.	1	.	.	.	.	.	.	1	.	.	1	.	1	.	1	1	.	1	1	2	2	3

Table 37(b). Mosses and some lichens in vegetation types with at least three relevés.

11. Survey of Alliances and lower syntaxonomical units

All. Saxifrago-Ranunculion nivalis
Suball. Luzulenion arcticae
Ass. Oxyrio-Trisetetum (A1-2, Z1)
- Var. of Saxifraga oppositifolia (A1)
- Var. of Potentilla crantzii (A2)
- Var. of Poa arctica (Z1)
- Ranunculus pygmaeus-Trisetum spicatum comm. (A3)

Ass. Phippsietum algidae-concinnae (Z2, B4)
Subass. typicum (Z2)
Subass. oxyrietosum digynae (B4)
- Typical variant (B4.2)
- Var. of Cerastium regelii (B4.3)
- Var. of Saxifraga oppositifolia (B4.1)
- Salix arctica-Saxifraga cernua comm. (B5)
- Subtype of Minuartia stricta (B5.2)
- Subtype of Taraxacum arcticum (B5.1)
- Subtype of Potentilla hyparctica (B5.3)
- Alopecurus alpinus-Saxifraga tenuis comm. (Z3)

Ass. Luzulo-Salicetum herbaceae (D9, Z4)
- Typical variant (D9)
- Var. of Oxyria digyna (Z4)

Ass. Koenigio-Saginetum intermediae (C7-8, Z5)
- Var. of Ranunculus glacialis (C7)
- Var. of Saxifraga hirculus (Z5)
- Var. of Saxifraga aizoides (C8)
- Eriophorum scheuchzeri stands (C6)

All. Caricion atrofusco-saxatilis
Ass. Saxifrago-Kobresietum simpliciusculae (E11-14, Z6)
- Var. of Carex saxatilis (E12)
- Subtype of Carex atrofusca (E12.2)
- Subtype of Carex capillaris (E12.1)
- Typical variant (E11)
- Var. of Carex bigelowii (E13)
- Var. of Juncus castaneus (Z6)
- Var. of Betula nana (E14)
- Subtype of Tofieldia pusilla (E14.2)
- Typical subtype (E14.1)

Ass. Arctagrostio-Eriophoretum tristis (F15-20, Z7-11)
Subass. typicum (Z7)
- Var. of Koenigia islandica (F18)
- Var. of Vaccinium microphyllum (F19)
- Var. of Saxifraga oppositifolia (F20.1)
- Subvar. of Cassiope tetragona (F20.2)
- Var. of Alopecurus alpinus (Z8)

Subass. betuletosum nanae (F15)
- Typical variant (F15.1)
- Var. of Carex rupestris (F15.2)

Subass. eriophoretosum scheuchzeri (Z9)
- Var. of Carex parallela (F17.2)
- Var. of Dupontia psilosantha (Z10)
- Var. of Alopecurus alpinus (Z11)
- Eriophorum triste-Rhododendron lapponicum comm. (F16)

Ass. Rhododendro-Vaccinietum microphylli (J24-25)
- Betula nana-Tofieldia pusilla comm. (J24)
- Subtype of Carex scirpoidea (J24.2b)
- Subtype of Kobresia myosuroides (J24.2a)
- Subtype of Equisetum variegatum (J24.2c-d)
- Subtype of Empetrum hermaphroditum (J24.1)
- Tofieldia coccinea-Carex bigelowii comm. (J25)
- Subtype of Carex rupestris (J25.1)
- Subtype of Vaccinium microphyllum (J25.2)
- Subtype of Euphrasia frigida (J25.3)

All. Dryadion integrifoliae
Ass. Saliceto-Cassiopetum tetragonae (G21, H22, J30, Z12-13)
Subass. pyroletosum grandiflorae (G21)
- Typical variant (G21.1)
- Var. of Empetrum hermaphroditum (G21.2)
- Var. of Pedicularis hirsuta (G21.3)
- Fragmentary communities (J30.1, H22)
- Intermediate type of Arctagrostis latifolia (Z12)
- Intermediate type of Empetrum hermaphroditum (Z13)

Subass. luzuletosum (J30.4)
- Var. of Dryas octopetala (J30.5)

Subass. typicum (J30.2-3)
  Arctostaphylos alpina-Betula nana comm. (I23)
Ass. Carici-Dryadetum integrifoliae
  Subass. caricetosum rupestris (J26-29, 31, K32)
  Typical variant (J26)
  Var. of Carex glacialis (J27)
  Var. inops (J28-29)
  Var. of Carex capillaris (J31)
  Var. of Hierochloë alpina (K32)

All. Veronico-Poion glaucae
Ass. Cerastio-Festucetum brachyphyllae (Z14-15, K34)
  Typical variant (Z14)
  Var. of Carex nardina (Z15)
  Var. of Draba glabella (K34)
  Draba glabella-Lesquerella arctica comm. (L39)
  Poa glauca comm. (M40)
  Pyrola grandiflora-Betula nana comm. (K33)
  Potentilla-Campanula uniflora comm. (K35)
Ass. Arabido holboellii-Caricetum supinae (L37-38, Z16)
  Typical variant (L37)
  Var. of Lesquerella arctica (L38)
  Var. of Draba arctica (Z16)
  Taraxacum phymatocarpum-Poa abbreviata comm. (N41)

All. Puccinellion phryganodis
Ass. Puccinellietum phryganodis
Ass. Caricetum subspathaceae

12. References

Aiken, S.G., Consaul, L.L. & Lefkovitch, L.P. 1995. *Festuca edlundiae* (Poaceae), a High Arctic, new species compared enzymatically and morphologically with similar *Festuca* species. - *Systematic Botany* 20.3: 374-392.

Barkman, J.J., Moravec, J. & Rauschert, S. 1986. Code of phytosociological nomenclature. - *Vegetatio* 67: 145-195.

Bay, C. 1992. *A phytogeographical study of the vascular plants of northern Greenland - north of 74° northern latitude.* - Meddelelser om Grønland, Bioscience 36. 102 pp.

- 1997. Floristic division and vegetation zonation of Greenland of relevance to a circumpolar arctic vegetation map. - *Journal of Vegetation Science*, in press.

- & Fredskild, B. 1990. *Botaniske undersøgelser i området mellem Fligely Fjord (74°50'N) og Nordmarken (77°30'N),* 1989. - Grønlands Hjemmestyre. Miljø- og Naturforvaltning. Teknisk rapport nr. 11. 56 pp.

- & Fredskild, B. 1991. *Botaniske undersøgelser i området mellem Germania Land (77°N) og Lambert Land (79°10'N), sommeren 1990.* - Teknisk rapport nr. 24. Grønlands Hjemmestyre, Miljø- og Naturforvaltning. 56 pp.

- & Holt, S. 1986. *Vegetationskortlægning af Jameson Land 1982-1986.* - Rapport. Grønlands Fiskeri- og Miljøundersøgelser. 40 pp. København.

Böcher, T.W. 1933. *Studies on the vegetation of the east coast of Greenland.* - Meddelelser om Grønland 104(4). 134 pp.

- 1954. *Oceanic and continental vegetational complexes in southwest Greenland.* - Meddelelser om Grønland 148(1). 336 pp.

- 1966. *Experimental and cytological studies on plant species. IX. Some arctic and montane Crucifers.* - Biol. Skr. Dan. Vid. Selsk. 14 (7). 74 pp.

- Fredskild, B., Holmen, K. & Jakobsen, K. 1978. *Grønlands Flora* (3. ed.) - Copenhagen. 312 pp.

Daniëls, F.J.A. 1982. *Vegetation of the Angmagssalik District, Southeast Greenland, IV. Shrub, dwarf shrub and terricolous lichens.* - Meddelelser om Grønland, Bioscience 10. 78 pp.

- 1994. *Vegetation classification in Greenland.* - J. Veg. Sci. 5: 781-790.

De Molenaar, J.G. 1976. *Vegetation of the Angmagssalik district, Southeast Greenland. II. Herb and snow-bed vegetation.* - Meddelelser om Grønland 198 (2). 266 pp.

Dierssen, K. 1996. *Vegetation Nordeuropas.* - Stuttgart. 838 pp.

Elkington, T.T. 1965. *Studies on the variation of the genus Dryas in Greenland.* - Meddelelser om Grønland 178 (1). 56 pp.

Elvebakk, A. 1985. Higher phytosociological syntaxa on Svalbard and their use in the subdivision of the Arctic. - *Nord. J. Bot.* 5: 273-284.

- 1994. A survey of plant associations and alliances from Svalbard. - *J. Veg. Sci.* 5: 791-802.

Evans, D.J.A. (ed.) 1995. *Cold Climate Landforms.* - Chichester and New York: John Wiley. 526 pp.

Fredskild, B. 1973. *Studies in the vegetational history of Greenland. Palaeobotanical investigations of some Holocene lake and bog deposits.* - Meddelelser om Grønland 198 (4). 247 pp.

- 1991. The genus *Betula* in Greenland - Holocene history, present distribution and synecology. - *Nord. J. Bot.* 11: 393-412.

- 1996. *Grønlands Botaniske Undersøgelse. Greenland Botanical Survey. 1995-1996.* - Botanisk Museum, København. 21 pp.

- & Bay, C. (ed.) 1993. *Grønlands Botaniske Undersøgelse. Greenland Botanical Survey. 1992.* - Botanisk Museum, København. 51 pp.

-, Bay, C. Holt, S. & Nielsen, B. 1986. *Grønlands Botaniske Undersøgelse. Greenland Botanical Survey. 1985.* - Botanisk Museum, København. 40 pp.

-, Bay, C., Feilberg, J. & Mogensen, G.S.

1992. *Grønlands Botaniske Undersøgelse. Greenland Botanical Survey. 1991.* - Botanisk Museum, København. 38 pp.

-, Mogensen, G.S. & Hansen, E.S. 1995. *Grønlands Botaniske Undersøgelse. Greenland Botanical Survey. 1994.* - Botanisk Museum, København. 20 pp.

Gelting, P. 1934. *Studies on the vascular plants of East Greenland between Franz Joseph Fjord and Dove Bay (Lat. 73°15'-76°20'N.)* - Meddelelser om Grønland 101 (2). 340 pp.

- 1937. *Studies on the food of the East Greenland ptarmigan especially in its relation to vegetation and snow-cover.* - Meddelelser om Grønland 116 (3). 196 pp.

Hadac, E. 1989. Notes on plant communities of Spitsbergen. - *Folia Geobot. Phytotax., Praha,* 24: 131-169.

Hansen, E.S. 1995. *Greenland Lichens.* - Copenhagen, Rhodos Publishers. 124 pp.

- 1996. Vertical distribution of lichens on the mountain, Aucellabjerg, Northeastern Greenland. - *Arctic and Alpine Research* 28: 111-117.

Hansen, H.M. 1930. *Studies on the vegetation of Iceland. - Botany of Iceland 3.* Copenhagen. 186 pp.

Hansen, K. 1969. *Analyses of soil profiles in dwarf-shrub vegetation in South Greenland.* - Meddelelser om Grønland 178 (5). 33 pp.

- & Jensen, J. 1974. *Edaphic conditions and plant-soil relationships on roadsides in Denmark.* - Dansk Botanisk Arkiv 28 (3). 143 pp.

Henriksen, N. 1994. Geology of North-East Greenland (75°-78°N) - the 1988-90 mapping project. - *Rapp. Grønlands geol. Unders.* 162: 5-16.

Holt, S. 1985. *Sneundersøgelser i relation til vegetation. Jameson Land 1984.* - Rapport. Grønlands Fiskeri- og Miljøundersøgelser, København. 76 pp.

Jackson, M.L. 1958. *Soil chemical analysis.* - Prentice-Hall, Englewood Cliffs N.J. 498 pp.

Jakobsen, B.H. 1988. Soil formation on the peninsula Tugtuligssuaq, Melville Bay, North West Greenland. - *Geografisk Tidsskrift* 88: 86-93.

- 1992a. Aspects of the genesis, geography and evolution of the soils of Greenland. - In: 1. *International Conference on Cryopedology, Puschino.* Russian Academy of Sciences. Proceedings: 71-84.

- 1992b. Preliminary studies of soils in North-East Grenland between 74° and 75° northern latitude. - *Geografisk Tidsskrift* 92: 111-115.

Holmen, K. 1952. *Cytological studies in the flora of Peary Land, North Greenland.* - Meddelelser om Grønland 128 (5): 40 pp.

- 1960. *The mosses of Peary Land, North Greenland.* - Meddelelser om Grønland 163 (2): 96 pp.

Holowaychuk, N. & Everett, K.R. 1972. *Soils of the Tasersiaq area, Greenland.* Meddelelser om Grønland 188 (6): 35 pp.

Lysgaard, L. 1969. *Foreløbig oversigt over Grønlands klima i perioderne 1921-50, 1951-60 og 1961-65.* - Det Danske Meteorologiske Institut. Meddelelser 21. 35 pp.

Nordhagen, R. 1936. *Versuch einer neuen Einteilung der subalpinen-alpinen Vegetation Norwegens.* - Bergens Museums Årbok 1936. Naturvidenskabelig rekke 7: 88 pp.

- 1943. *Sikilsdalen og Norges Fjellbeiter.* - Bergens Museums Skrifter 22: 607 pp.

Raup, H.M. 1965. *The structure and development of turf hummocks in the Mesters Vig district, Northeast Greenland.* - Meddelelser om Grønland 166.3. 112 pp.

- 1969a. *The relation of the vascular flora to some factors of site in the Mesters Vig district, Northeast Greenland.* - Meddelelser om Grønland 176.5. 80 pp.

- 1969b. *Observations on the relation of vegetation to mass-vasting processes in the Mesters Vig district, Northeast Greenland.* - Meddelelser om Grønland 176.6. 216 pp.

Schuster, R.M. & Damsholt, K. 1974. *The Hepaticae of West Greenland from ca. 66°N to 72°N.* - Meddelelser om Grønland 199.1: 373 pp.

Schwarzenbach, F.H. 1951. Ökologische Beiträge zur quartären Florengeschichte Ostgrönlands. - *Bericht über das Geobotanische Forschungsinstitut Rübel in Zürich für das Jahr 1950: 44-66.*

-1960. Die arktische Steppe in den Trockengebieten Ost- und Nordgrönlands. - *Bericht des Geobotanischen Institutes der Eidg. Tech. Hochschule, Stiftung Rübel* 31: 42-64.

-1961. *Botanische Beobachtungen in der Nunatakkerzone Ostgrönlands zwischen 74° und 75°N.Br.* - Meddelelser om Grønland 163.5. 172 pp.

Seidenfaden, G. & Sørensen, T. 1937. *The vascular plants of Northeast Greenland from 74°30' to 79°00' N. Lat.* - Meddelelser om Grønland 101.4. 215 pp.

Sørensen, T. 1933. *The vascular plants of East Greenland from 71°00' to 73°30' N. Lat.* - Meddelelser om Grønland 101.3. 177 pp.

- 1935. *Bodenformen und Pflanzendecke in Nordostgrönland.* - Meddelelser om Grønland 93.4. 69 pp. [English version: Ground form and plant cover in Northeast Greenland. See: D.J.A. Evans 1955.]

- 1941. *Temperature relations and phenology of the Northeast Greenland flowering plants.* - Meddelelser om Grønland 125.9. 305 pp.

- 1942. *Untersuchungen über die Therophytengesellschaften auf den Isländischen Lehmflächen ("Flags").* - Biol. Skr. Vid. Selsk. II.2: 31 pp.

- 1945. *Summary of the botanical investigations in N.E.Greenland.* - Meddelelser om Grønland 144.3: 48 pp.

- 1948. *A method of establishing groups of equal amplitude in plant sociology based on similarity of species content and its application to analyses of the vegetation on danish commons.* - Biol. Skr. Vid. Selsk. V.4: 34 pp.

- s.a. *Summarisk Redegørelse for et ufuldendt Arbejde over Nordøstgrønlands Plantesamfund (72°-74°N.) baseret paa Vegetationsanalyser, foretaget 1931-33.* - In the files of Botanical Museum, Copenhagen.

Tedrow, J.C.F. 1970. *Soil investigations in Inglefield Land, Greenland.* - Meddelelser om Grønland 188.3. 93 pp.

Ugolini, F.C. 1966a. *Soils of the Mesters Vig district, Northeast Greenland. 1. The Arctic Brown and related soils.* - Meddelelser om Grønland 176.1. 23 pp.

- 1966b. *Soils of the Mesters Vig district, Northeast Greenland. 2. Exclusive of Arctic Brown and Podzol-like soils.* - Meddelelser om Grønland 176.2. 27 pp.